U0313290

The Macat Library
世界思想宝库钥匙丛书

解析詹姆斯·E. 拉伍洛克
《盖娅：地球生命的新视野》

AN ANALYSIS OF

JAMES E. LOVELOCK'S

GAIA: A NEW LOOK AT LIFE ON EARTH

Mohammad Shamsudduha ◎ 著
庄稼 ◎ 译

目　录

CONTENTS

引 言

要点

- 詹姆斯 · E. 拉伍洛克，世界上最重要的独立科学家之一，杰出的科普书籍作家，1919 年出生于英格兰东南部的一个工人阶级 * 家庭。
- 拉伍洛克的《盖娅：地球生命的新视野》认为，这颗行星上的生物与大气、海洋和岩石相互作用，形成可自我调节的稳定生物圈 *，生命在其中生长旺盛（此处的“生物圈”是指星球上被生物占据的部分）。
- 拉伍洛克具有争议性的盖娅假说 *——该假说认为，地球是有生命的超个体 *（就像生态群落一样由许多不同个体或生物体组成的一个实体）——促进了地球系统科学 * 这门全新学科的发展。

詹姆斯 · E. 拉伍洛克其人

詹姆斯 · E. 拉伍洛克，《盖娅：地球生命的新视野》（1979）的作者，1919 年出生于英格兰东南部的一个工人阶级家庭。[1] 父亲托马斯 · A. 拉伍洛克 * 是一位艺术商，而母亲内莉 · A. 伊丽莎白 * 则在市政府担任私人秘书。[2] 双亲都未曾接受过多少正规的教育。内莉因为生计 13 岁就已辍学。托马斯在孩童时没上过学，一直到后来读专科学校才学会阅读和书写。[3] 拉伍洛克的父母也许正是因为自己的经历，十分重视教育，总是鼓励他们的儿子上学——但由于家庭经济困难，拉伍洛克到了年龄却无法读大学。他选择了一家化学咨询公司的培训职位，同时参加夜校课程。

后来，拉伍洛克获得奖学金并进入曼彻斯特大学学习，[4] 于

1941 年从化学专业 * 毕业。1948 年，他获得医学博士学位。1959 年，他又获得伦敦大学生物物理学（将物理学应用于研究生物体的学科）博士学位。[5]

拉伍洛克大部分时间作为独立科学家而进行科研，英国《卫报》把他称为"世界知名作家和演说家"。[6] 其声名主要源于他激进的盖娅假说：地球是一个由许多不同个体或生物体组成的实体——类似于蜜蜂或蚂蚁群落的有生命的超个体。英国哲学家玛丽·米奇利 * 将拉伍洛克比作 19 世纪进化 * 论的奠基人查尔斯·达尔文 *："虽然对于细节把握精确，他却没有因为关注细节而忽略更广阔、要求更高的整体背景。他着眼大处。就像达尔文一样，他更喜欢在机构的轨道之外从事研究。从 1963 年起，他一直依靠发明和咨询服务自给自足。"[7]

《盖娅：地球生命的新视野》的主要内容

从 1969 年起，拉伍洛克的一些学术论文已开始讨论这一想法，[8] 而他真正着手撰写《盖娅：地球生命的新视野》是在 1974 年，当时他住在爱尔兰西部一处自然风景秀美的地方。在这本处女作里，他用浅显易懂的语言将思想传递给广大读者。他向普罗大众介绍了极具争议性的观点，即我们的星球是一个可自我调节的生命实体，并以古希腊大地女神"盖娅"为之命名。

对拉伍洛克而言，人类并非地球的独特属性，而是更宽泛的生物群体中的一部分。拉伍洛克参考了人文主义 * 的政治定位与哲学定位，同时强调了人类事务的重要性。他写道："我越来越多地从她（盖娅）的视角看待一切，就像慢慢地脱下旧外套一样，慢慢地消除曾经忠诚的人文主义基督教信仰，不再认为人类共同福利是唯一重要的事。"[9]

拉伍洛克的假说 * 认为，生机盎然的地球是一个超个体——由许多其他生命体所组成的生命体，生命体之间相互作用，同时还与空气、海洋和这颗星球表面的岩石相互作用。这一相互作用的系统使地球适宜生存。拉伍洛克首次提出生命体控制着其所在的无生命环境，该思想被证明是极具争议性的。

拉伍洛克清晰地表达其思想，如同叙述一个由章节串联起来的故事一般。由于写作风格平铺直叙，缺少证据，还运用了神话和诗歌，这部作品在 1979 年发表之初就受到科学界非常严厉的批评[10]。来自不同学科的科学家，其中包括生物学 * 和地质学 * 的科学家，都对其观点持严肃的保留意见。（生物学研究生物体；地质学研究岩石等地球物质的组成与结构。）

拉伍洛克的盖娅假说，其实质在于认为整个地球表面，包括所有生物体，是一个可自我调节的整体。它根据需要而改变自身，以保持其物理、化学和生物环境之间的平衡，从而维持生命并帮助其随时间推移而进化。通过自我调节，地球为生物维持稳定状态的持久趋势，又称为“动态平衡”*。这颗星球的动态平衡便是“盖娅假说”的基础——在科学界与哲学界均为激烈辩论的话题。[11]

由于拉伍洛克的盖娅假说与持续发生的气候变化 *——通常理解为人类行为所导致的地球气象模式或平均温度的大规模、长期变化——的讨论和争议相关，所以盖娅假说与科学领域一直保持密切联系。在科学数据库“科学指引”* 中搜索短语“盖娅假说”，会得到 373 篇科学论文和 146 部著作。此外，在谷歌搜索引擎上键入短语“盖娅假说”，可以找到 400 000 条结果。

《盖娅：地球生命的新视野》的学术价值

从 20 世纪 80 年代起，拉伍洛克的盖娅假说不断影响着一批又一批学者、科学家、政治家和普通民众。最初，许多科学家反对可自我调节的、有生命的地球这一思想，就像德国科学家阿尔弗雷德·魏格纳*的理论，它认为地球上的陆地在该星球表面上漂移并移动位置（“大陆漂移说”*）；以及英国博物学家查尔斯·达尔文的自然选择进化论，它解释了环境、适应和遗传特性如何导致新物种的产生。两者现今都被公认为事实。[12]

最初，拉伍洛克的假说受到地质学家、进化生物学家*（根据进化论研究地球生物的科学家）以及行星科学家*（对行星和卫星进行研究的科学家）的严厉批评。例如，进化生物学家理查德·道金斯*认为这些生物体不能作为一个整体来控制其所在环境。[13]

尽管受到广泛批评，但是许多科学家还是认为拉伍洛克的观点值得更深入探讨。于是乎拉伍洛克的盖娅假说在80年代后期开始得到更多支持和关注。[14]1985年，马萨诸塞大学举办了首届盖娅假说公开研讨会，主题为“地球是一个活的生物体吗？”。1988年，美国地球物理学联合会*在加利福尼亚州圣地亚哥召开了首届查普曼会议，主题便是盖娅假说。[15]“改变是潜移默化的”，拉伍洛克于1994年这样写道。此时，以“自我调节的地球”为主题的科学会议正在英格兰的牛津举行。2000年，美国地球物理学联合会又在西班牙瓦伦西亚举办了第二届会议。[16]2006年，以盖娅假说为主题的又一场国际会议在弗吉尼亚州的乔治梅森大学召开。[17]

为应对最初的批评，拉伍洛克与他的主要合作者——美国地球科学家林恩·马古利斯*以及英国海洋与大气科学家安德鲁·沃森*将盖娅假说发展为可以经科学测试的理论（“地球科学”是科学的一个分支，利用地质学、物理学和化学来研究地球定义属性）。尽管“盖娅假说”和“盖娅理论”通常可互换使用，但拉伍洛克认为两者分别代表着不同含义。对他而言，只有在科学依据支持之下，假说方才成为理论——在《盖娅时代》（1995）一书中，拉伍洛克这样写道："现如今，我们可以开始把盖娅视作理论，而不仅仅是假说。"[18]

"即使是要认识大气层这一地球最基本的特征，"拉伍洛克指出，"成为地球物理学家*还不够，还需要化学和生物学知识"[19]（地球物理学研究我们这颗行星各种不同的引力、地磁、电以及地震现象等特性）。盖娅假说的核心思想涉及范围广泛多样，无法立刻判定适用哪一门主要的科学学科。因此，拉伍洛克建议这样宽泛的主题应在全新的跨学科的科学分支（即采用各种不同科学领域的探索目标和方式来进行研究的一门科学分支）内进行讨论。

最终，盖娅假说创造了地球生理学*这门全新的学科——又被称为地球系统科学，研究地球生命与地球本身之间的相互作用。许多世界顶尖的学术机构，包括加利福尼亚州的斯坦福大学，都已成立地球系统科学系，研究地球的海洋、陆地以及大气所组成的一体化系统。

---

1. 伊恩·欧文："詹姆斯·拉伍洛克：绿色运动支持者"，载《独立报》2005年12月3日，登录日期2013年10月10日，http://www.independent.co.uk/news/people/profiles/james-lovelock-the-greenman-517953.html。
2. 詹姆斯·E.拉伍洛克：《致敬盖娅：独立科学家的生命》（修订版），伦敦：纪念品出版社，2014年，第1页。
3. 詹姆斯·E.拉伍洛克：《消失的盖娅：最终警告》，伦敦：企鹅出版集团，2010

年，第 206 页。
4. 罗宾·麦凯："盖娅的战士"，载《绿色生活方式杂志》2007 年 7—8 月，第 60—62 页。
5. "简历"，詹姆斯·拉伍洛克官方网站，登录日期 2013 年 12 月 29 日，http://www.jameslovelock.org/page2.html。
6. 彼得·福布斯："吉姆会处理"，载《卫报》2009 年 2 月 21 日，登录日期 2013 年 12 月 23 日，http://www.theguardian.com/culture/2009/feb/21/james-lovelock-gaia-book-review。
7. 玛丽·米奇利："伟大的思想家——詹姆斯·拉伍洛克"，载《新政治家》2003 年 7 月 14 日。
8. 詹姆斯·E. 拉伍洛克和 C. E. 吉芬："行星大气层：与生命存在有关的组成或其他变化"，载《宇航科学发展》1969 年第 25 期，第 179—193 页。
9. 参见詹姆斯·E. 拉伍洛克："前言"，《盖娅：地球生命的新视野》，詹姆斯·E. 拉伍洛克（修订版），牛津：牛津大学出版社，2000 年，第 ix 页。
10. W. 福特·杜利特尔："大自然真的是慈母吗？"，载《共同进化季刊》1981 年第 29 期，第 58—65 页。
11. 理查德·R. 华莱士和布莱恩·G. 诺顿："盖娅理论的政策意义"，载《生态经济学》1992 年第 6 期，第 103 页。
12. 马丁·奥格尔："盖娅理论：21 世纪的科学模型与隐喻"，载《门槛杂志》2009 年第 1 期，第 99—106 页。
13. 参见理查德·道金斯：《延伸表现型：作为物竞天择最小单位的遗传因子》，牛津：牛津大学出版社，1982 年，第 234—236 页。
14. 拉伍洛克：《盖娅：地球生命的新视野》，第 xii 页。
15. 艾瑞克·G. 考夫曼："盖娅论战：美国地球物理学联合会的查普曼会议"，载《美国地球物理学联合会学报 Eos》1989 年第 69 期，第 763—764 页。
16. 布伦特·F. 鲍曼：《检验盖娅假说的可行性》，学士论文，詹姆斯·麦迪逊大学，1998 年。
17. "乔治梅森大学举办盖娅理论会议"，阿灵顿县，登录日期 2013 年 12 月 27 日，http://www.gaiatheory.org/2006-conference/。
18. 詹姆斯·E. 拉伍洛克：《盖娅时代：地球传记》（修订版），牛津：牛津大学出版社，1995 年，第 44 页。
19. 詹姆斯·E. 拉伍洛克："地球生理学，盖娅的科学"，载《地球物理学评论》1989 年第 27 期，第 222 页。

第一部分：学术渊源

1 作者生平与历史背景

要点

- 《盖娅：地球生命的新视野》对可自我调节的地球这一概念加以解释，这个概念是对维持地球上生命的相关科学辩论的全新且重要的贡献。“可自我调节”用以描述地球生物、空气、水和岩石之间的相互作用，从而使地球适合生命生存。
- 拉伍洛克的生涯中大多数时间为独立科学家，对他而言，自由思考对发展盖娅假说至关重要。其假说认为，“地球表面、大气和海洋的物理与化学条件，是由生命存在本身主动创造的，以使其适合生命存在。”[1]
- 拉伍洛克曾在美国国家航空航天局*（美国国家航空航天局负责美国民用太空项目）短暂任职，在加利福尼亚州的喷气推进实验室里进行无人维京火星探测计划*。这段经历让他对研究地球与其他星球上的生命产生了浓厚兴趣。

为何要读这部著作?

詹姆斯·E. 拉伍洛克的《盖娅：地球生命的新视野》（1979）是当代科学界有关生命和行星变化进程讨论的最畅销书籍之一。作品描写了一个可自我调节的、有生命的地球，并以古希腊大地女神“盖娅”为之命名。拉伍洛克激进的盖娅假说受到进化生物学家、化学家（具备物质性质和反应方面的专业知识）以及行星科学家（研究行星和卫星）的严厉抨击。即便如此，仍有学者提出其假说已发展为可经科学测试的理论——盖娅理论*。[2]

在过去的三十年间，无数篇学术论文、著作和数次国际会议都探讨了盖娅假说。[3] 世界各地的科学家和研究人员在 2001 年通过的《阿姆斯特丹全球变化宣言》* 中承认："地球是一个由物理、化学、生物和人类等各部分组成的可自我调节的单一系统。"[4] 虽然《宣言》未提及盖娅，但这正是盖娅假说的核心思想。

政府间气候变化专门委员会 * 开发了地球系统模型 * 以预测未来气候状况，这是盖娅理论的主要成就。[5] 政府间气候变化专门委员会是联合国 * 领导的跨政府组织。地球系统模型通过研究大气、海洋、陆地、冰层和生物圈之间的相互作用来评估在不同条件下地区和全球气候的状况。

> "二战加速了发明创造，这使得人文和科学领域繁荣发展。倘若我们不是如此天性好战的动物，那么我们原本可以更加建设性地应用这些新的知识。我们原本可以把从宇宙观察地球作为当务之急，建造人造卫星观察陆地、空气和海洋，及时捕捉到迫在眉睫的全球变暖问题。然而恰恰相反，我们建造了导弹。"
>
> —— 詹姆斯 · E. 拉伍洛克：《通往未来的艰难之旅》

作者生平

詹姆斯 · E. 拉伍洛克 1919 年 7 月 26 日出生于英格兰莱奇沃斯花园城的一个工人阶级家庭。[6] 父亲托马斯 · A. 拉伍洛克对画画很感兴趣；[7] 母亲内莉 · A. 伊丽莎白是一位秘书。在其自传《致敬盖娅》中，拉伍洛克这样写道："对我母亲最沉重的打击是她获得了一个很难得到的奖学金……可以到文法学校（一种公立的重点学校）就读，然而她却必须放弃，因为家里需要十三岁的她挣钱养家糊口。她未能接受启蒙教育……把时间都花费在泡菜厂给广口瓶贴标签。"[8] 母亲总是怀着“工人阶级的善意，”拉伍洛克继续写道，“她毫不犹豫地坚信教育。她下定决心让我就读文法学校，并且越快越好。被剥夺接受‘良好教育’机会的她不打算让我也受同样的苦。”[9]尽管拉伍洛克并不喜欢学校生活，但他决心要成为一名科学家。

拉伍洛克在伦敦的国家医学研究所*开启了他的职业生涯。他在此工作了近20年，期间仅有为数不多的几次暂歇。因为不想永久性地在一家教育或研究机构工作，他于60年代初成为一名独立科学家。此后，他再也不曾加入任何重要大学或研究机构，而是依靠发明和出版物的收入独立从事科研。他曾写道：“独立的乐趣之一在于：尽管顾客不同，但其需求在很大程度上却是共同分享的。为一家机构完成的工作经常与为另一家机构所做的工作相得益彰，好比为美国国家航空航天局所做的工作与为壳牌（跨国石油公司）所做的工作那样。”[10]拉伍洛克认为，这种独立性对他发展激进的盖娅假说帮助巨大。

创作背景

拉伍洛克的童年、教育背景、实践经历和他的许多研究合作者，都极大影响了他对地球生命与众不同的思考方式。从童年起，他就对大自然和科学产生了浓厚的兴趣。《绿色生活方式杂志》[11]记录了他对科学的热情源于伦敦的科学博物馆、自然历史博物馆，以及《世界大战》（1898）的作者英国作家赫·乔·威尔斯*与《地球到月球》（1865）的作者法国作家儒勒·凡尔纳*的科幻小说。拉伍洛克的原生家庭也极大地影响了他的思考方式以及对自然的关爱。最大的影响来源于他爱戴的父亲，父亲终其一生热爱并关心自然。[12]拉伍洛克在其作品中描述说，他父亲曾说过，地球表面的每一个生物都有其存在目的，它们汇聚在一起便组成了更大的生态系统 *（在某一特定地区内发现的全部生物体所组成的生物系统），在这个系统中，物理环境与生命体相互作用。

拉伍洛克对盖娅的探索始于20世纪60年代初，当时他正在美国国家航空航天局（NASA）的喷气推进实验室进行“维京火星探测计划”，探测火星生命。这个无人太空探索计划基于这样一种理论，即在火星上发现的任何生命的证据都与地球生命的证据相类似。拉伍洛克提出，只要简单地对火星大气进行测试，就能知道火星上是否有生命。很快，他对究竟生命是什么的课题产生了兴趣，并积极致力于在行星进程框架内深入研究地球大气层、海洋组成和生物体 * 的作用。

拉伍洛克的可自我调节的地球这一思想产生之初，全球气候变化（地球天气模式以及地区与全球气温的长期变化）或生物多样性 *（地球生命的多种多样）的概念尚未引起公众重视。其盖娅假说解释了温室气体 *（留住太阳能量的气体）如何保持大气温暖而舒适，从而保护了地球早期的生命形态。他写道，“栖息地被破坏与大气中温室气体不断增加的危险，在20世纪70、80年代看来是如此遥远而微不足道。”[13]

---

1. 詹姆斯·E. 拉伍洛克：《盖娅：地球生命的新视野》（修订版），牛津：牛津大学出版社，2000年，第144页。
2. 詹姆斯·W. 基什内尔：“盖娅假说：猜想与反驳”，载《气候变化》2003年第58期，第21页。

3. “盖娅假说”，环境网站，登录日期2013年12月23日，http://www.environment.gen.tr/gaia/70-gaia-hypothesis.html。
4. 詹姆斯·E. 拉伍洛克：“有生命的地球”，载《自然》2003年第426期，第769—770页。
5. 詹姆斯·E. 拉伍洛克：《通往未来的艰难之旅》，伦敦：企鹅出版集团，2015年，第94页。
6. 伊恩·欧文：“詹姆斯·E. 拉伍洛克：绿色运动支持者”，载《独立报》2005年12月3日，登录日期2013年10月10日，http://www.independent.co.uk/news/people/profiles/james-lovelock-the-greenman-517953.html。
7. 詹姆斯·E. 拉伍洛克：《致敬盖娅：独立科学家的生命》(修订版)，伦敦：纪念品出版社，2014年，第7—10页。
8. 拉伍洛克：《致敬盖娅》，第7—37页。
9. 拉伍洛克：《致敬盖娅》，第15—16页。
10. 拉伍洛克：《致敬盖娅》，第282页。
11. 罗宾·麦凯：“盖娅的战士”，载《绿色生活方式杂志》2007年7—8月，第60—62页。
12. 拉伍洛克：《致敬盖娅》，第8页。
13. 参见詹姆斯·E. 拉伍洛克：“前言”，《盖娅：地球生命的新视野》，第viii页。

2 学术背景

要点

- 当拉伍洛克供职于美国国家航空航天局时，他测试火星生命的构想备受嘲讽。科学界最初直接忽视了他的盖娅假说（假说认为地球表面与其所供养的生命组成了一个可自我调节的整体）。
- 拉伍洛克着手发展盖娅假说时，并未意识到在他之前已有一些科学家简要地讨论过可自我调节的地球这一思想（此处的“自我调节”是指地球的有机特征与非有机特征之间的相互作用，使其适合生命舒适地生存）。
- 拉伍洛克与他的一名学生一起创建了环境的数学模型，并称之为“雏菊世界”*，用以测试其构想。

著作语境

詹姆斯 · E. 拉伍洛克的作品《盖娅：地球生命的新视野》的创新贡献在于理解了环境与生物体对地球气候的影响作用。20 世纪 50 年代末至 60 年代是太空探索 * 的新时期。1957 年，苏联 * 发射了人造地球卫星 *，这是绕地球飞行的第一颗人造卫星 *。[1] 鉴于美国与苏联的竞争由来已久，于是在 1961 年，约翰 · 肯尼迪 * 总统开始通过美国国家航空航天局来发展美国的太空计划。他设定的目标是在十年内实现人类登陆月球并安全返航。[2]

美国国家航空航天局还计划向火星发射宇宙飞船，找寻生命的踪迹。那时几乎还没有对地球之外的生命做过研究。生物学家 *

（研究生物体的科学家）设计实验，想再现火星的环境，但这些试验只能在地球上完成。供职于美国国家航空航天局的拉伍洛克提出质疑，认为在地球上做试验对于发现火星生命是不可行的。作为通过对火星的土壤取样来探测是否存在生命的备选方案，他提出可以对火星的大气环境取样。他的想法遭到美国国家航空航天局的驳回和嘲笑。[3]

同样，他在 20 世纪 60 年代参加会议时第一次介绍了其盖娅假说，但也未能引起科学界的关注。可自我调节的地球这一思想相当复杂，不能用单一学科领域来框定。一直到 1979 年《盖娅》一书出版，学术界才对其假说产生兴趣。

> “我原以为能在科学文献中找到生命作为物理过程的全面定义，人们可以把发现生命的试验建立在这样的定义之上。然而，我却惊讶地发现有关生命本质的书写是如此贫乏。”
>
> —— 詹姆斯 · E. 拉伍洛克：《盖娅：地球生命的新视野》

学科概览

尽管在拉伍洛克提出其思想之前，正式的学术领域从未讨论过可自我调节的地球这一假说，[4] 但是就像拉伍洛克所写的那样：“狭义地说，地球是有生命的这一思想的历史或许与人类的历史一样长。”[5] 过去，数位伟大的科学家曾把这颗星球看作生命体。1785 年，著名的苏格兰地质学家詹姆士 · 赫顿 * 形容地球是个可自我调节的系统。赫顿认为地球就像一个生物，地球的养分（即食物）在土壤和植物间循环，水资源从海洋到陆地来回移动，就好比人体的血液循环。[6] 在《盖娅时代》(1995 年)一书中，拉伍洛克这样记录道："詹姆士·赫顿无疑在地质学领域影响深远，但经历 19 世纪简化论 * 大潮，大家都忘记抑或是否定了其有生命的地球的思想" [7]("简化论"是指把一个复杂的事物看作其各组成部分的总和，而不考虑每一部分之间可能产生相互作用)。20 世纪初，俄国科学家弗拉基米尔·伊·维尔纳茨基 * 探讨了表面覆盖生命的活着的地球这一概念，并普及了"生物圈"这一术语，用以描述生命体存在的地区。[8]

尽管如此，拉伍洛克在形成自己的假说时并不知晓这些思想，其假说比前人更为深入地发展了可自我调节的地球这一思想。许多科学家都认为拉伍洛克的盖娅假说与颇具影响力的英国博物学家查尔斯·达尔文的进化论相矛盾。进化论认为生物体不断进化，而无生命的环境不会进化。但后来，拉伍洛克证明了生物体会与其无生命的环境产生相互作用，以形成最适宜其生存的条件。[9]

学术渊源

拉伍洛克的童年经历、与双亲的关系、教育和工作经历，以及与其他科学家的合作，都对他思考地球生命的独特视角产生了重要的影响。他也受到数位研究前辈的影响。他在自传中说，他深受英国裔美国动物学家乔治·伊·哈钦森 * 有关地球生物化学 * 研究工作的影响(动物学家对动物进行科学研究；生物化学研究生命体的化学过程)。哈钦森研究生物体与其环境的相互作用，从化学活动的角度把地球称作可自我调节的实体。[10] 拉伍洛克还记载了他的化学家朋友悉尼·埃普顿 *。1975 年两人在科普期刊《新科学人》发表合著文章，第一次真正激发了公众对盖娅假说的兴趣。[11]

可自我调节的地球的概念以及盖娅假说发展中的成型和影响逐步扩大，离不开拉伍洛克的主要合作者——地球科学家林恩·马古利斯、哲学家戴恩·希契科克*与悉尼·埃普顿。拉伍洛克与马古利斯早在70年代初就已合作研究盖娅假说。拉伍洛克在书中把两人的伙伴关系称为“最令人受益的科学合作”，该合作促成了他们两人关于这一主题的首篇科学论文[12]。马古利斯不仅是研究搭档，更是他的好朋友，她很相信拉伍洛克，还为他争取到了研究盖娅假说并出版第二部著作《盖娅时代》（1995年）的资金。[13]后来，来自英国的博士候选人安德鲁·沃森加入拉伍洛克的研究。沃森与拉伍洛克一起创建了“雏菊世界”这样一个虚构的类地行星数学模型[14]，以证明生物体实际上可以控制它们所生存的大气与气候。

---

1. 詹姆斯·E.拉伍洛克：《盖娅时代：地球传记》（修订版），牛津：牛津大学出版社，1995年，第4页。
2. “太空计划”，约翰·F.肯尼迪总统图书馆与博物馆，登录日期2016年1月8日，http://www.jfklibrary.org/JFK/JFK-in-History/Space-Program.aspx。
3. 詹姆斯·E.拉伍洛克：《致敬盖娅：独立科学家的生命》（修订版），伦敦：纪念品出版社，2014年，第250页。
4. 拉伍洛克：《盖娅时代》，第3—14页。
5. 拉伍洛克：《盖娅时代》，第9页。
6. 拉伍洛克：《盖娅时代》，第9页。
7. 拉伍洛克：《盖娅时代》，第9页。
8. 拉伍洛克：《盖娅时代》，第10页。

9. 拉伍洛克：《盖娅时代》，第 41—61 页。
10. 拉伍洛克：《盖娅时代》，第 263 页。
11. 詹姆斯·E. 拉伍洛克和悉尼·埃普顿："探索盖娅"，载《新科学人》1975 年第 65 卷第 935 期，第 304—309 页。
12. 詹姆斯·E. 拉伍洛克和林恩·马古利斯："生物圈的环境平衡：盖娅假说"，载《地球》1974 年第 26 卷第 1—2 期，第 1—10 页。
13. 拉伍洛克：《致敬盖娅》，第 369 页。
14. 安德鲁·詹·沃森和詹姆斯·E. 拉伍洛克："全球环境的生物平衡：雏菊世界的寓言"，载《地球》1983 年第 35B 期，第 286—289 页。

3 主导命题

要点

- 20 世纪 60 年代中期，当拉伍洛克构思可自我调节的地球这一想法时，美国国家航空航天局的研究人员和学者们还在致力于将仅在地球上测试过的生物试验应用于在地球以外的行星上发现生命。
- 拉伍洛克的盖娅假说认为地球上的每一个生物体都与其所在环境紧密相连，其相互作用的总和形成了适宜生存的条件。
- 以前的科学家曾附带提出过地球是可自我调节的、有生命的系统，但拉伍洛克倾其一生努力证明这一观点。

核心问题

阅读詹姆斯·拉伍洛克的《盖娅：地球生命的新视野》时，我们必须参考作者对于“什么是生命？”以及“如何认识生命？”这两个命题的思考。

虽然这两个核心问题本质简单，但对于拉伍洛克发展生命进化以及地球作为有生命的单一实体的思想，确实是独创且至关重要的。他花费多年精力，将其激进而具有开创性的思想发展为盖娅假说，提出地球上的每一个生物体——从微小的病毒到庞大的鲸——都与其对应的环境紧密相连，从而构成单一的实体。通过控制地球大气和气候，这些与环境相连接的实体能够维持适宜生存的条件。

拉伍洛克假说的构建离不开他 20 世纪 60 年代中期在美国国家

航空航天局喷气推进实验室的工作经历。在那里，他被阿波罗 * 载人航天计划的宇航员所拍摄的地球照片深深吸引。他曾写道，这些照片引导他自上而下而非自下而上地观察地球表面。在从事探索其他星球生命的这段工作期间，有生命的地球这一想法进入了他的脑海。他认为一定存在着全球规模的调节系统，使地球适宜生命生存，而火星和金星这些近邻却没有生命迹象。[1] 美国国家航空航天局的科学家们对火星土壤进行试验，以发现生命。拉伍洛克对其有效性提出质疑，并建议研究火星的大气成分，以检测是否存在生命。[2]

> "……思索火星生命，为我们提供了考虑地球生命的新颖角度，引导我们构思一种全新的抑或是重振一种古老的有关地球及其生物圈之间关系的概念。"
>
> —— 詹姆斯 · E. 拉伍洛克:《盖娅：地球生命的新视野》

参与者

当拉伍洛克在 60 年代中期构思有生命的地球这一想法时，有关该主题的辩论并不多。在两大意识形态对立的大国——苏联与美国——竞争全球军事与文化主导地位之时，太空探索风头正劲。美国国家航空航天局的生物学家和土壤学家致力于宇宙飞船的生命探测实验。但这些实验均在地球表面进行。生物化学家万斯 · 小山 * 便是其中之一，他坚持认为应收集火星土壤，以检验生命是否存在。那时，还没有任何传统学科讨论过"生命"的定义以及生命能否控制地球表面的生存环境。[3]

拉伍洛克理解的盖娅，即有生命的地球，与生命的概念紧密相连。确实，要理解盖娅就必须理解生命。在着手彻底研究生命及其对地球环境的意义时，拉伍洛克发现几位科学家前辈已提出地球是活的这一思想。1785 年，被称为地球科学之父的苏格兰地质学家詹姆士 · 赫顿提出，地球是一个超个体——是由栖居其中的生物相互作用而组成并由此被定义的有生命的实体。19 世纪末，英格兰生物学家托玛斯 · H. 赫胥黎 * 提出，地球是一个有生命的、可自我调节的系统。而俄国地球化学家 * 弗拉基米尔 · 伊 · 维尔纳茨基认为，如同积极的地质力量一样，生命也在塑造并改变地球。[4] 但是，这些早期的科学家并没有在最初的思想基础上更进一步。而拉伍洛克采取崭新的研究方式探索该思想，并将它视为毕生的工作。他构建了盖娅假说，解释了地球如何维持适宜生命生存的条件。这种不断变化中的稳定性又被称为“动态平衡”。

当代论战

尽管有生命的地球这一思想为前人所构思，但主流科学家并未接受它，且除了几部早期科学出版物之外该思想也不为人所知。例如，乌克兰哲学家、独立科学家叶夫格拉夫 · M. 科罗连科 * 在 19 世纪末宣称地球是活的生命体。[5] 但当拉伍洛克本人构思有生命的地球这一想法并开始提出相同的命题时，他并不知道从前有过这些表述或作品。当时没有正在进行的讨论或文章能够直接用来解答可自我调节的地球这一思想的相关核心命题。

这一思想的主题宽泛，拉伍洛克需要建议与合作，才能将假说发展为理论，从而能够进行可测试的科学预测。[6] 许多科学家直接批评了该思想，但也有些科学家愿意与他合作，尤其是戴恩 · 希契科克、悉尼 · 埃普顿以及林恩 · 马古利斯，他们积极地帮助拉伍洛克发展盖娅假说。

1. 詹姆斯·E. 拉伍洛克：《盖娅：地球生命的新视野》（修订版），牛津：牛津大学出版社，2000 年，第 1—29 页。
2. 詹姆斯·E. 拉伍洛克：《致敬盖娅：独立科学家的生命》（修订版），伦敦：纪念品出版社，2014 年，第 242—243 页。
3. 詹姆斯·E. 拉伍洛克：《盖娅时代：地球传记》（修订版），牛津：牛津大学出版社，1995 年，第 15—20 页。
4. 参见克里斯平·蒂克尔："前言"，《盖娅的复仇：地球为何在还击——我们怎样才能拯救人类》，詹姆斯·E. 拉伍洛克著，伦敦：企鹅出版集团，2007 年，第 xiv 页。
5. 拉伍洛克：《盖娅时代》，第 8—10 页。
6. 拉伍洛克：《盖娅时代》，第 41—61 页。

4 作者贡献

要点

- 尽管拉伍洛克撰写本书的主要目的在于向公众传递可自我调节的、有生命的地球这一核心思想，但他知道科学家们可能会读到其盖娅假说。
- 拉伍洛克具有争议性的盖娅假说改变了人们对于地球生命的看法，并最终创造了地球生理学或地球系统科学这一全新的学术领域——该学科建立在如下的基本原则之上：地球是一个相互作用的体系，该体系中生命起着至关重要的作用。
- 从可自我调节的地球这一思想逐渐发展为盖娅理论，历时十年以上。通过与不同学科的科学家讨论，并在得到一些密切合作者的帮助之后，该思想发展成为盖娅理论。

作者目标

詹姆斯·拉伍洛克于 1974 年起开始创作《盖娅：地球生命的新视野》一书。他相信，包括人类在内的所有生物组成了一个生态群落，不知不觉地让地球适宜生存。他甚至开始认为人类和其他生物体一样，没有特别的权力，只承担盖娅群落的义务。[1] 拉伍洛克想与公众分享他的想法，而不仅仅是针对当时还无法接受其争议性的科学家们。

盖娅思想的灵感最早形成于 1965 年，当时拉伍洛克正在美国国家航空航天局工作。1967 年，他在国际期刊《伊卡洛斯》* 上第一次将该思想公之于众，并在 1971 年发表了题为“大气环境下的盖娅”的演讲，将这一思想展示给他的科学家同行们。[2]但是有声望的主流期刊，例如《科学》与《自然》，尚未准备好接受有关盖娅假说的文章。“通过分析大气来探测行星生命的想法，”拉伍洛克写道，“对审稿的传统天文学家 * 与生物学家而言一定不能被接受。”[3]

1979 年，拉伍洛克终于发表了《盖娅》，收录了他到当时为止的全部思想。他知道其思想的重要性，并且意识到需要一套具体方案，把假说变为可试验的科学理论。在接下来的十年中，他与许多科学家进行讨论，但只有少数支持他的思想并与他合作。拉伍洛克想向科学家同行们证明对地球应该自上而下进行观察研究，而非自下而上。（他坚持认为）只有在太空中观察地球，并与它死气沉沉的邻居们作比较，才能看出我们的蓝色星球是充满生机的。

> “地球是一种生物体，能够调节气候与组成，使其适合栖居其中的生物，孕育这一思想的环境是让人崇敬的科学殿堂。1965 年某个下午，我突然得到了这一灵感，当时我正在加利福尼亚州的喷气推进实验室工作。”
>
> —— 詹姆斯 · E. 拉伍洛克：《盖娅：地球生命的新视野》

研究方法

《盖娅：地球生命的新视野》叙述了在地球这颗行星上发现生命的故事。作品以描述美国国家航空航天局在太阳系的其他行星上寻找生命的太空计划而拉开序幕。拉伍洛克创造性地提出了他对在火星或金星这样遥远的世界探测生命的核心问题的解决方法，即行星的大气层可以提供其表面是否存在生命的必要证据，而无需对土壤进行检测。这一具有争议性的思想背离了当时的科学正统，拉伍洛克也知道没有证明就不会有人相信。

1965 年至 1975 年间，他尽可能地搜集了支持信息，将其个人思想发展为可行的假说。他认为描写其假说的最佳方法是写一个关于发现的故事。拉伍洛克用一系列相互串联的章节，叙述了早期地球的大气环境，以及生命是如何从微小的单细胞生物进化为像人类一样复杂的形态。拉伍洛克解释了生物体是如何与其所在的无生命环境相互影响，并构成一个可自我调节的实体，也就是他笔下的"盖娅"。在他之前，还没有哪个作家如此深入地分析过盖娅之中生命体与无生命体之间错综复杂的互补关系。

时代贡献

虽然可自我调节的、有生命的地球这一核心概念并不新鲜，但拉伍洛克采取的研究方法却具有原创性。那时还没有人把生命进化与地球进化作为一个整体来进行过研究。当拉伍洛克着手详细研究关于生命及其与所在环境关系的现有文献时，他完全找不到相关的内容。但他偶然发现一些旧笔记和文献曾提出过地球是活的。例如，18 世纪的地质学家詹姆士·赫顿认为，地球就像一个动物。一个世纪之后，奥地利地质学家爱德华·修斯*引入了"生物圈"这个词，后来又被俄国地球化学家弗拉基米尔·伊·维尔纳茨基继续扩展，指出其为地球上产生能量维持生命的区域。[4]尽管拉伍洛克对前人的相关思想一无所知，但后来他在第二部著作《盖娅时代》(1995）中引用了这些先驱者的思想，向他们致敬。

---

1. 詹姆斯·E. 拉伍洛克:《盖娅:地球生命的新视野》(修订版),牛津:牛津大学出版社,2000 年,第 ix 页。
2. 詹姆斯·E. 拉伍洛克:《盖娅时代:地球传记》(修订版),牛津:牛津大学出版社,1995 年,第 8 页。
3. 詹姆斯·E. 拉伍洛克:《致敬盖娅:独立科学家的生命》(修订版),伦敦:纪念品出版社,2014 年,第 250 页。
4. 拉伍洛克:《盖娅时代》,第 8—10 页。

第二部分：学术思想

5 思想主脉

要点

- 盖娅假说的关键部分是平衡而稳定的生存条件，又被称为动态平衡，是生命体与其所在环境之间相互作用形成的结果。
- 控制论反馈机制 *——一种根据变化做出调整的自动控制系统——维持着动态平衡。
- 拉伍洛克曾警告说，如果对大气环境和地球表面产生过多改变，那么人类可能会破坏维持盖娅的反馈机制。

核心主题

詹姆斯·E. 拉伍洛克的《盖娅：地球生命的新视野》(1979) 讲述的是一个寻找生命的故事。故事的核心便是盖娅假说。假说认为，地球不断维持其物理、化学和生物环境之间的相互作用，使生物体得以生存并随着时间逐渐进化。这种相对的稳定性，又称为动态平衡，是理解盖娅假说的根本。

拉伍洛克在《盖娅》一书中写道，地球通过一个自动系统来保持动态平衡，这种系统又被称为控制论反馈机制。[1] 该系统根据变化作出调整，控制着地球大气层和海洋的温度与组成。书中提到，约 35 亿年前，地球初期的大气层刚刚形成，盖娅——有生命的地球——使大气环境适应生物体，保护了这颗星球上的生命。大气层还保护地球免受宇宙辐射 *——来自太空的能量波——的破坏，也免受陨石（来自太空的岩石）的轰炸。[2]

作品最终的主题是盖娅内的可持续生存。拉伍洛克描写了人类与盖娅之间的关系，并警告我们地球内部的反馈机制失灵会导致可怕的后果。如果人类对地球表面与大气环境进行过多改变，就会导致机制失灵。书中还记叙了人类比以往任何时候燃烧更多的石油、天然气和煤炭，从而急剧增加了大气层的二氧化碳 * 气体。这导致了全球变暖 *（全球气温上升），也就是温室效应 *：大气层中二氧化碳和其他温室气体的浓度增加，从而留住更多太阳的热量。[3]

> “地球表面温度主动维持在适宜复杂实体盖娅的水平，并且在盖娅存在的大多数时间内一直保持如此。”
>
> —— 詹姆斯 · E. 拉伍洛克：《盖娅：地球生命的新视野》

思想探究

在《盖娅》一书中，拉伍洛克解释了地球大气的组成在几百万年中是如何变化的。以氮和氧为主的各种大气气体共同构成了动态稳定的环境（差不多就是平衡），即使环境不断改变，生命也得以生存。拉伍洛克认为地球与其邻近的行星并不相同。[4] 火星与金星的大气中主要是二氧化碳气体，而几乎没有维持生命的氧气。与此相反，地球大气含 21% 左右的氧气，对动植物而言至关重要。拉伍洛克用一个概念模型来展示一个类地行星如果没有生命且达到化学平衡态，那么只可能存在微量的维持生命所需的氧气。[5] 但是地球大气的氧气含量在过去的两亿年里都保持在 21% 左右。[6] 持续稳定的氧浓度表明地球表面存在着一套主动控制系统。

拉伍洛克将该控制系统称为控制论反馈机制。他以电烤炉、熨斗和室内取暖器这样一些家用电器为例解释这一概念。这些家电都配备了恒温器 *，它通过打开或关闭家电就可以将其控制到理想的温度。如果温度超过需求，反馈机制就会关闭家电，使其维持特定的温度。与之类似，人体也有反馈机制，可维持体内温度，所以我们的身体机能能够使我们生存下去。[7]拉伍洛克强调，地球的控制论反馈机制是全球范围的，动植物都有能力调节地球气候。[8]

在最后两章里，拉伍洛克转而讨论了盖娅的可持续性生存主题。他告诫读者，如果人类继续改变地球表面和大气环境，那么行星脆弱的动态平衡系统——它通过数十亿年发展而成——将最终瓦解，人类将面临严峻后果。拉伍洛克表达了极大的恐慌，因为生命与盖娅的共存状态危在旦夕，而人类为了自身着想，不断改变着地球地貌及其脆弱的大气环境。他在书中写道："温室气体产生率的改变，将引发全球范围的微小变化"[9]，从而影响盖娅的动态稳定状态，最终可能威胁地球生命的生存。[10]

语言表述

拉伍洛克的《盖娅》一书受众是一般读者，语言简单，非专业人士也能轻松理解可自我调节、有生命的地球这一古怪而陌生的思想："我写这本书，只需要词典的帮助。"[11]科普杂志《新科学人》评论认为："拉伍洛克文笔优美。既具创新性又写得很好的图书无疑是额外的收获。"[12]

拉伍洛克获得了成功。简单而有效的方法最终为他的科普作家身份正名。但由于文风朴实，缺乏证据，主题具有争议性，不少科学家对作品提出了强烈批评。拉伍洛克决定用古希腊大地女神"盖娅"来命名可自我调节的地球，突显了作品非科学的本质[13]（"盖娅"是《蝇王》的作者英国小说家威廉·戈尔丁建议使用的名字，他与拉伍洛克居住在同一个村庄）。[14]

尽管存在问题，但拉伍洛克还是使其思想得到了科学家的认可，所以他后来表示，绝不后悔选择了这一名字。

---

1. 参见詹姆斯·E. 拉伍洛克："控制论"，《盖娅：地球生命的新视野》，詹姆斯·E. 拉伍洛克修订，牛津：牛津大学出版社，2000 年，第 44—58 页。
2. 拉伍洛克：《盖娅：地球生命的新视野》，第 59—77 页。
3. 拉伍洛克：《盖娅：地球生命的新视野》，第 100—132 页。
4. 詹姆斯·E. 拉伍洛克：《致敬盖娅：独立科学家的生命》（修订版），伦敦：纪念品出版社，2014 年，第 244 页。
5. 拉伍洛克：《盖娅：地球生命的新视野》，第 30—46 页。
6. 詹姆斯·E. 拉伍洛克：《盖娅时代》（修订版），牛津：牛津大学出版社，1995 年，第 124 页。
7. 拉伍洛克：《盖娅：地球生命的新视野》，第 49 页。
8. 拉伍洛克：《盖娅：地球生命的新视野》，第 58 页。
9. 拉伍洛克：《盖娅：地球生命的新视野》，第 113 页。
10. 詹姆斯·E. 拉伍洛克：《通往未来的艰难之旅》，伦敦：企鹅出版集团，2015 年，第 75—111 页。
11. 布伦特·F. 鲍曼：《检验盖娅假说的可行性》，学士论文，詹姆斯·麦迪逊大学，1998 年，第 8 页。
12. 肯尼思·梅兰比：《与地球母亲一起生活》，载《新科学人》1979 年第 84 期，第 41 页。
13. 托比·蒂勒尔：《论盖娅：生命与地球关系的批判性调查》，普林斯顿：普林斯顿大学出版社，2013 年，第 2 页。
14. 拉伍洛克：《致敬盖娅》，第 255 页。

6 思想支脉

要点

- 在地球早期，太阳的热量远不如今天这般强烈，温室气体吸收太阳热量，使当时的地球表面保持温暖而舒适，维持脆弱的生命。
- 可自我调节机制使大气层的含氧量和海洋的含盐量保持相对稳定。
- 起名“盖娅”并讨论有生命的地球，在一开始就转移了科学界对盖娅假说的注意力，直到后来才重新获得关注。

其他思想

詹姆斯·E. 拉伍洛克的《盖娅：地球生命的新视野》一书中，一些次要思想对其发展具有开创性的盖娅假说至关重要：

- 如何控制地球大气环境中的各种气体浓度？
- 化学反应——不同化学物质化合的作用——对维持地球生命的重要性在哪里？
- 温室气体对于保护早期地球的作用是什么？
- 为什么大海的含盐量不会更高一些？又是什么控制了海水的盐浓度？

以提问形式来陈述的这些次要思想帮助拉伍洛克解释并扩展主题。例如，他解释了化学反应的恒定能量供应是如何维持生命的。两种或两种以上的化学物质化合，形成不同的化学物质，这就是化学反应。如果一颗行星处于化学意义上的动态稳定（平衡）状态，那么就不发生任何反应或产生任何能量。[1] 拉伍洛克解释说：“这样的世界不存在任何能量源：不降雨，没有波浪和潮汐，产生能量的化学反应就不可能发生。”[2]火星和金星不存在化学反应，也就不产生能量——离我们最近的这些行星邻居确实了无生气。[3]

> “人类当然是盖娅的关键进展之一，但我们在她生命中出现得很晚，现在开始讨论我们自身与她的关系似乎并不合适。”
>
> —— 詹姆斯 · E. 拉伍洛克：《盖娅：地球生命的新视野》

思想探究

地球生命很有可能发源于简单的单细胞，生存在宇宙撞击和活跃的放射性 * 之下的不稳定条件下[4]（原子衰变并释放辐射 * 形成了放射性——大约就是能量波）。由于大气缺少氧气，地球表面暴露在肉眼无法看见的太阳紫外线 * 辐射之下。早期地球的大气环境中主要是二氧化碳气体和氨气 *（氮氢化合物）。拉伍洛克提出，35 亿年前，地球发出的热量比今天少 25%，是这些温室气体使这颗星球保持温暖。[5]拉伍洛克认为，太阳热量减少 25% 意味着地表平均温度远低于 0 摄氏度，导致地球表面被冰雪覆盖。然而，根据地质记录，我们知道这一时期的地球气候并非完全不适合生命生存。[6]因此，随着太阳在过去的 35 亿年里逐渐变强，盖娅在相对稳定的气候下维持着地球生命。

地球大气环境主要由氮和氧组成，还有少量的二氧化碳和其他微量气体。这些气体对调节地球温度和气候发挥了重要作用，并受生物区 *——环境中动植物生物体的控制。比如说，生物区的积极控制能使大气中的含氧量保持恒定。绿植和藻类 *（一种类似植物的水生生物）通过光合作用*，利用太阳光产生氧气，在过程中生成营养物质并散发氧气。这增加了大气中维持生命的氧气——而含氧量保持约21%不变。

那么盖娅如何控制地球大气环境，以保持含氧量适宜生命生存？控制氧气的关键是另一种气体——甲烷*，一种碳氢化合物，主要由单细胞生物体——细菌*产生。虽然大气中仅含微量甲烷，但它却是控制含氧量的关键因素。拉伍洛克把甲烷的作用比作人体血液中的葡萄糖（一种糖类）。[7] 葡萄糖提供了人体细胞所需的能量，所以对细胞健康以及由此带来的身体健康而言，保持较稳定的血糖水平是必不可少的。与之类似，通过与氧气反应，甲烷可以形成二氧化碳和水，从而控制着大气含氧量。

拉伍洛克解释了如何平衡全世界各大洋的含盐量。雨水与河水溶解了岩石里的盐分，并穿越陆地将其带入海洋，但海洋的含盐量在很长时间里都恒定地保持在约3.4%。[8] 拉伍洛克认为，根据化石记录，历史上海洋的含盐量不可能超过6%。如果含盐量更高，我们今天见到的海洋生物可能进化得大相径庭。[9] 拉伍洛克提出，由于海洋的含盐量稳定，因此必然存在着某种机制，能够减少海洋的某些盐分。[10] 他认为："过剩的盐会在蒸发率*高且（盐分）单向从海洋中流入的区域堆积，例如浅海湾、陆地环绕的泻湖以及孤立的狭长海域，并形成蒸发岩*（沉积物）。"[11]

被忽视之处

许多科学家误解了拉伍洛克的盖娅假说。拉伍洛克在修订版中写道，作品"受众并非严格意义上的科学家。如果他们不顾我的警告而阅读此书，他们会发现其观点太过激进，甚至从科学角度上看是不正确的。”[12] 但科学家们还是把作品当作科学文本来阅读，反应也和拉伍洛克描述的如出一辙。进化生物学家绝不同意有生命的动物可以通过与其无生命的环境相互影响从而改变环境这一思想。其他科学家认为盖娅假说是目的论 *（符合自然是有目的的这一理念），是目标导引的 *（基于重复分析寻找特定答案）。他们提出，生物体如果可自我调节，那么势必具有预见和规划的能力，而这并无可能。[13]

拉伍洛克对此作出评价：“他们把盖娅视为（超越科学的）元科学，就像宗教信仰，他们出于根深蒂固的唯物主义信仰，会加以抵制。”[14] 拉伍洛克的两位美国的支持者，生物学家斯蒂芬 · 施奈德 * 和环境科学家佩内洛普 · J. 波士顿 *，在他们的作品《研究盖娅的科学家》（1993）[15] 中写道，盖娅假说错误地吸引了神学家 * 的关注，他们通常通过经文来研究宗教思想。“有生命的地球”这一措辞以及“盖娅”这一命名，可能使科学界的关注点偏离了假说及其含义的严肃分析。

---

1. 参见詹姆斯 · E. 拉伍洛克：“认识盖娅”，《盖娅：地球生命的新视野》，詹姆斯 · E. 拉伍洛克修订，牛津：牛津大学出版社，2000 年，第 32 页。
2. 拉伍洛克：《盖娅：地球生命的新视野》，第 33 页。
3. 拉伍洛克：《盖娅：地球生命的新视野》，第 30—43 页。
4. 参见詹姆斯 · E. 拉伍洛克：“最初时刻”，《盖娅：地球生命的新视野》，詹姆斯 · E. 拉伍洛克修订，牛津：牛津大学出版社，2000 年，第 15 页。
5. 拉伍洛克：《盖娅：地球生命的新视野》，第 18 页。

6. 拉伍洛克：《盖娅：地球生命的新视野》，第 18 页。
7. 拉伍洛克：《盖娅：地球生命的新视野》，第 67 页。
8. 凯特·拉维琉斯："完美的和谐"，载《卫报》2008 年 4 月 28 日，登录日期 2013 年 12 月 30 日，http://www.theguardian.com/science/2008/apr/28/scienceofclimatechange.biodiversity。
9. 拉伍洛克：《盖娅：地球生命的新视野》，第 86 页。
10. 詹姆斯·E. 拉伍洛克：《盖娅时代：地球传记》（修订版），牛津：牛津大学出版社，1995 年，第 99—107 页。
11. 拉伍洛克：《盖娅：地球生命的新视野》，第 91 页。
12. 拉伍洛克：《盖娅：地球生命的新视野》，第 xii 页。
13. 詹姆斯·E. 拉伍洛克：《致敬盖娅：独立科学家的生命》（修订版），伦敦：纪念品出版社，2014 年，第 264 页。
14. 拉伍洛克：《盖娅：地球生命的新视野》，第 xii 页。
15. 斯蒂芬·H. 施奈德和佩内洛普·J. 波士顿编：《研究盖娅的科学家》，坎布里奇：麻省理工学院出版社，1993 年，第 433 页。

7 历史成就

要点

- 拉伍洛克的《盖娅：地球生命的新视野》主要面向普通读者，因为他相信科学界不会重视这部作品，也确实如此。
- 他对损害地球的可自我调节机制的危害提出警告，与当今对气候变化的担忧直接相关。
- 盖娅假说帮助建立了被称为地球系统科学的跨学科研究领域。

观点评价

詹姆斯·拉伍洛克的第一部作品《盖娅：地球生命的新视野》在1979年发表之初毁誉参半。后来，拉伍洛克在其自传中写道："它的发表彻底改变了我的一生，通过邮箱寄给我的信件从淅淅沥沥到倾盆大雨一般，在那之后更是络绎不绝。"[1]对盖娅的兴趣最主要还是来自普通读者、哲学家以及宗教领袖，只有三分之一的信件来自科学家。[2]

虽然撰写这本书的目的不是为了向专家们提供阅读的纯科学文本，但拉伍洛克仍然预料一些科学家会阅读该作品。他们确实阅读了——不少人并不接受盖娅假说。该书一经出版，拉伍洛克的科学界同僚们就开始询问，为什么用一本书来报告盖娅假说的思想，而不是发表在经同行评审的科学期刊上。[3]他回应道：20世纪70年代初，当他首次发表讨论盖娅的科学论文时，主要来自进化生物学家的批评太过强烈，他认为诸如《科学》或《自然》这样有声望的期刊的编辑不会接受此类文章。[4]

但情况在90年代得以改观，即便是要在有声望的期刊上发表盖娅主题的论文，也变得容易起来。拉伍洛克这样评论道：生物学家们早期的抨击和一部分科学期刊编辑的保守，帮助他发现了假说的潜能，以不断推进后来的科学辩论与发现。[5]

> “如今大多数科学家似乎都接受了盖娅理论，并将之应用在其研究中，但他们仍然拒绝使用‘盖娅’这个名字，而更愿意称之为地球系统科学，或地球生理学。”
>
> —— 詹姆斯·E. 拉伍洛克：《盖娅：地球生命的新视野》

当时的成就

拉伍洛克的盖娅假说诞生于 20 世纪 60 年代中期，当时美国与苏联正在进行激烈的太空探索竞赛，也就是说太空占据了关注的焦点。他在美国国家航空航天局的工作是进行火星生命探测的科学试验，引导他思考地球为何与火星和金星这些邻居不一样。

自作品于 1979 年发表以来，科学界一直就盖娅假说进行着激烈的辩论。辩论的结果之一是开发了“雏菊世界”这一数学模型。[6] 生态学家利用该模型测试生物多样性 *（某地区生物的丰富性）以及生态系统（环境、栖居其中的生物体，以及两者之间的相互影响）稳定性对于维持健康生存环境的作用。[7]

在这本书里，拉伍洛克还解释了过度改变地球表面和大气环境是如何导致地球脆弱的可自我调节系统崩溃的。盖娅假说在 20 世纪 70 年代预测全球变化时，没有证据提供证明。现如今，人们普遍认可例如二氧化碳等大气中的温室气体发生改变，会在全球范围内影响气候。[8] 这些变化被称为“人为因素”*，意思是由人类活动引起的。拉伍洛克在 30 多年前对人类活动会通过温室效应 * 导致全球变暖的危险提出过警告。盖娅假说与我们今天对气候变化的担忧直接相关。

局限性

《盖娅》出版于 1979 年，语言易懂不晦涩，深受普通读者喜爱，但一些科学家并不接受，尤其是进化生物学家。[9] 除了科学性的分歧之外，作品采用讲故事的写作方式，还使人联想到神话人物，所以科学家认为无法严肃对待该作品。[10] 拉伍洛克承认："《盖娅》一书写的是假设，笔调轻描淡写——就像一幅粗糙的铅笔素描，试着从不同的视角刻画地球之貌。"[11] 他还如此回应科学家们的批评："写这本书时，我们只不过刚刚瞥见这颗星球的实质，我把它写作为一个发现的故事。"[12]

《盖娅》的另一限制因素在于其思想，可自我调节的地球这一概念相当宽泛，不能纳入例如地质学、生物学或物理学等任何单一的传统学科领域。[13] 在当时，跨学科的研究方式还不普遍。如今，盖娅假说已经从美国进化生物学家史蒂芬·杰·古尔德 * 理解的"并非原理，而是隐喻 *"[14]，变为一门新兴的跨学科研究领域的核心：地球系统科学。拉伍洛克的盖娅假说并不局限于某一时间或地点，还启发了不少其他学科，包括生态学 *（生物学的分支，研究生物体之间以及生物体与其物理环境之间如何相互联系）、海洋生物学 *（研究海洋生命）以及气候科学 *。[15]

---

1. 詹姆斯·E. 拉伍洛克：《致敬盖娅：独立科学家的生命》（修订版），伦敦：纪念品出版社，2014 年，第 264 页。

2. 拉伍洛克：《致敬盖娅》，第 264 页。
3. 詹姆斯 · E. 拉伍洛克：《盖娅时代：地球传记》(修订版)，牛津：牛津大学出版社，1995 年，第 xiii—xxii 页。
4. 拉伍洛克：《盖娅时代》，第 xv 页。
5. 拉伍洛克：《盖娅时代》，第 xiv—xv 页。
6. 安德鲁 · 詹 · 沃森和詹姆斯 · E. 拉伍洛克："全球环境的生物平衡：雏菊世界的寓言"，载《地球》1983 年第 35B 期，第 286—289 页。
7. 詹姆斯 · E. 拉伍洛克：《消失的盖娅：最终警告》，伦敦：企鹅出版集团，2010 年，第 115 页。
8. 詹姆斯 · E. 拉伍洛克：《盖娅：地球生命的新视野》(修订版)，牛津：牛津大学出版社，2000 年，第 113 页。
9. 拉伍洛克：《致敬盖娅》，第 264 页。
10. 拉伍洛克：《盖娅：地球生命的新视野》，第 xi 页。
11. 拉伍洛克：《盖娅时代》，第 11 页。
12. 拉伍洛克：《盖娅：地球生命的新视野》，第 viii 页。
13. 拉伍洛克：《盖娅：地球生命的新视野》，第 xii—xiii 页。
14. 史蒂芬 · 杰 · 古尔德："克鲁泡特金不是疯子"，载《博物学》1997 年第 106 期，第 12—21 页。
15. 拉伍洛克：《致敬盖娅》，第 242—279 页。

8 著作地位

要点

- 拉伍洛克致力于盖娅假说的研究长达半个世纪之久。在 20 世纪 60 年代中期的一场学术会议上，他第一次提出了可自我调节的地球这一思想。
- 《盖娅》一书发表后，拉伍洛克和他的合作者又投入了十年的精力，以可试验且更为科学严密为目标，将该概念发展为更广义的盖娅理论。
- 由于盖娅并不完全适应任何现有的学科，地球系统科学这一全新的学术领域因此诞生。

定位

《盖娅：地球生命的新视野》的作者詹姆斯·E. 拉伍洛克十分珍惜可以遵循自己想法开展研究的这份自由。正因为此，他一直没有与任何一家研究或学术机构保持太长时间的联系。这也反映了他把科学探索看作与世界直接接触的一种形式。[1]

作为独立科学家和发明家，拉伍洛克创造了许多科学设备，其中电子捕获探测器 *——用于探测并测量类似氯氟烃 * 的大气的气体——是其最重要的发明。拉伍洛克写道："电子捕获探测器无疑是最有价值的商品，它帮助我在各种学科之中不断探索盖娅，还帮助我真正地走遍地球。"[2] 尽管这一发明与盖娅假说并无关联，但它帮助拉伍洛克走进了美国国家航空航天局的太空研究实验室。假说的最初构想在此孕育，往后更带来了名望和全球的认可。

第一本有关盖娅的作品于1979年发表之后，拉伍洛克的职业生涯发生了些许转变。他不再专心于科学发明，而是投入更多时间和精力到盖娅的研究，撰写著作和科学期刊论文。他的研究重点就是其盖娅假说的进一步发展。十多年来，他咨询不同学科的科学家，试图将盖娅假说从可自我调节的地球这一相对简单的思想不断进行扩展。通过合作者的协助，他成功地把假说发展为理论*，能够解释人类对环境的干预，例如排放二氧化碳和甲烷这些温室气体，将如何影响地球的动态平衡状况，危及脆弱的气候系统。[3]此处的“动态平衡”是指倾向于平衡的状态，无论环境如何改变，生命体对自身保持控制。

拉伍洛克在职业生涯里共撰写了超过200篇有关盖娅的科学论文，还发表了10部著作。《盖娅：地球生命的新视野》是他的第一部作品，也是最出名的一部。

> “作为一名科学家，我完全服从科学规则，这也就是为什么我彻底修改了我的第二部作品——《盖娅时代》，希望科学家们能接受它。作为一个人，我所在的自然历史世界更温和，可以充满诗意地表达思想，任何感兴趣的人都能理解，这也就是为什么（《盖娅》）这本书基本保持原样。”
>
> —— 詹姆斯·E. 拉伍洛克：《盖娅：地球生命的新视野》

整合

拉伍洛克职业生涯的第一部分主要投入于医学研究。[4]45岁的他供职于美国国家航空航天局时，构思了可自我调节的地球的主要思想，这些思想于20世纪60年代中期发展为盖娅假说。之后，该假说成了拉伍洛克研究和发表的主要领域。在本书写作过程中，年逾九旬的他仍活跃在盖娅研究中。关于盖娅的最新一部作品《通往未来的艰难之旅》于 2015 年出版。[5]

拉伍洛克写道："探索盖娅的路自始至终都是战斗。"[6] 然而，尽管各科学圈都对该假说进行着激烈的辩论，即便是最执着的批评者——美国地球科学家 * 詹姆士 · 基什内尔 *——也提到该假说确实激发了其他许多科学家形成属于自己的假说。[7]（地球科学是利用地质学、化学、生物学以及气候学等多种学科领域，来研究我们这颗星球的深度历史和系统功能。）与几位科学家的合作，尤其是与林恩 · 马古利斯和安德鲁 · 沃森的合作，使该假说真正发展为可测试的理论。

任何科学理论的价值都可由其预测的精准度来判断。[8] 拉伍洛克的盖娅理论做出了 10 条预测，包括火星不存在生命，以及微生物和生物多样性对调节行星气候起关键作用。迄今为止，其中 8 项预测已得到验证，或至少为科学家们普遍接受。[9] 环境科学家和盖娅评论家托比 · 蒂勒尔在其作品《论盖娅》（2013）中写道："无论从海水化学成分方面的生物控制角度，还是从大气气体组成的角度，拉伍洛克（认为生命改变了地球）这一观点都是正确的。"[10]

盖娅理论借鉴了许多学科——从天体生物学 *（研究地球之外的生命的科学分支）到生态学，还对天体物理学 *（研究恒星、星系、行星等物理特性的天文学分支）、生物学和地球科学等许多领域的科学研究和发现作出重要贡献。最后，盖娅理论的概念与应用引发了地球系统科学这一新兴跨学科学术领域的发展。[11]

意义

在过去的 40 年里，拉伍洛克的盖娅假说影响了许多学者、科学家和政治家。例如，捷克前总统瓦茨拉夫·哈维尔曾说过："根据盖娅假说，我们是一个更大整体的组成部分。我们的命运不仅取决于我们为自己做了什么，而且取决于我们为盖娅这个整体做了什么。"[12]但要科学界接受该假说并不容易。

虽然拉伍洛克在他的领域已经因为其科学发明而出名，但盖娅假说真正让他名扬天下。《盖娅：地球生命的新视野》够大胆也够激进，足以引发科学界持续数十年的争论，也证明了作品在当代科学中的重要意义。

在科学数据库"科学指引"* 中搜索短语"盖娅假说"，会得到146部著作和373篇科学论文。今天，相关主题的科学论文还在源源不断地发表。2006年，伦敦地质学会，世界上最古老的地质专业团体，授予了拉伍洛克最高荣誉——沃拉斯顿奖*，以表彰其对地球研究的终身贡献。[13]

---

1. 约翰·格雷："詹姆斯·E. 拉伍洛克：处变不惊之人"，载《新政治家》2013年3月27日，登录日期2013年12月21日，http://www.newstatesman.com/culture/culture/2013/03/james-lovelock-man-all-seasons。
2. 参见詹姆斯·E. 拉伍洛克："前言"，《盖娅：地球生命的新视野》，詹姆斯·E. 拉伍洛克修订，牛津：牛津大学出版社，2000年，第xvii页。
3. 詹姆斯·E. 拉伍洛克：《消失的盖娅：最终警告》，伦敦：企鹅出版集团，2010年，第23—45页。
4. 詹姆斯·E. 拉伍洛克：《致敬盖娅：独立科学家的生命》(修订版)，伦敦：纪念品出版社，2014年，第69—104页。
5. 詹姆斯·E. 拉伍洛克：《通往未来的艰难之旅》，伦敦：企鹅出版集团，2015年。

6. 拉伍洛克:《致敬盖娅》, 第 278 页。
7. 詹姆斯 · W. 基什内尔:"盖娅假说:猜想与反驳", 载《气候变化》2003 年第 58 期, 第 21 页。
8. 拉伍洛克:《消失的盖娅》, 第 116—117 页。
9. 拉伍洛克:《消失的盖娅》, 第 116 页。
10. 托比 · 蒂勒尔:《论盖娅:生命与地球关系的批判性调查》, 普林斯顿:普林斯顿大学出版社, 2013 年, 第 202 页。
11. 詹姆斯 · E. 拉伍洛克:"地球生理学, 盖娅的科学", 载《地球物理学评论》1989 年第 27 期, 第 215—222 页。
12. 拉伍洛克:《盖娅:地球生命的新视野》, 第 x 页。
13. 拉伍洛克:《艰难之旅》, 第 77 页。此处引用提及的获奖年份(2003)是错误的。奖项颁发于 2006 年。参见"沃拉斯顿奖嘉奖令", 詹姆斯 · 拉伍洛克官方网站, 登录日期 2016 年 3 月 4 日, http://www.jameslovelock.org/page7.html。

第三部分：学术影响

9 最初反响

要点

- 最重要的批评是抨击拉伍洛克的盖娅假说“无法进行科学测试”，简单来说就是目的论（基于自然是有目的的这一理念），是目标导引（基于重复分析寻找特定答案）。
- 拉伍洛克针对盖娅假说的批评，撰写了其第二部著作《盖娅时代》，专门迎合科学家标准。
- 有证据显示生物体确实可调节其栖居环境，足以将该假说变为可测试的理论，从而使科学界越来越倾向于接受盖娅原理。

批评

詹姆斯·E. 拉伍洛克的盖娅假说最初受到激烈的反对。最强烈的抨击来自哈佛大学地球化学家海因里希·霍兰*。他否认生物体控制并调节地球大气与气候（地球化学研究地球固态物质的化学成分及其化学变化）。霍兰认为气候受地球化学和地球物理进程控制，而生命并未主动参与其中。[1]

美国生物学家福特·杜利特尔*也提出了负面评价。他于1981年在《共同进化季刊》上发表了对盖娅假说的评论。杜利特尔认为，没有证据显示单个生物体能如假说所提出的那样提供控制论反馈机制——一种物理、生物或社会反应，可影响系统的持续活动或生产力。[2]他得出的结论是，盖娅假说没有任何解释机制，是不符合科学的理论。

另一位进化生物学家，英国人理查德·道金斯，在《延伸表现型：作为物竞天择最小单位的遗传因子》（1982）[3] 中也批评了盖娅假说。他认为生物体缺乏远见和计划，因而无法聚合成一个统一的整体。他与杜利特尔的评论相附和，也不同意控制论反馈机制能够稳定系统。与拉伍洛克提出的观念不同，道金斯认为，物竞天择绝不可能导致全球范围的利他主义 *（无私的行为）。[4] 道金斯认为盖娅假说是目的论，是目标导引。

美国进化生物学家史蒂芬·杰·古尔德将盖娅假说形容为地球进程的隐喻式描写——不按字面意思理解的修辞手法。[5] 与此同时，地球科学家詹姆士·基什内尔于 1989 年这样写道："盖娅有多种不同的外观，混合了事实、理论、隐喻和不切实际的想法。"[6] 对他而言，盖娅假说认为地球环境条件为适应生物体的需要而以某种方式进行改变，这一点无法通过科学验证，甚至是误导性的 [7]。正常理解的自然选择 *，即进化论最基本的原则之一，是生物体更好地适应其不断变化的环境的过程。

尽管拉伍洛克声明作品受众是普通读者，但未使用科学专业语言这一点受到了科学家们的批评。[8] 其他人表示作品充其量不过是表达宗教信仰或是现代精神幻想。[9]

> "批评家们对待科学一丝不苟，对他们而言，任何联系到神话或讲故事的成分都使之沦为伪科学……我试过……重写我的第二本书——《盖娅时代》，在迎合科学家标准的同时，维持（《盖娅》）这本书不变（向科学家们作出证明）。"
>
> —— 詹姆斯·E. 拉伍洛克：《盖娅：地球生命的新视野》

回应

拉伍洛克用不同的方式回应了科学家同行们的批评，表示批评者们根本没读懂，抑或是用错误的标准来评价《盖娅》一书。他对认为盖娅假说不可测试这一方面的批评进行了积极应对。拉伍洛克的回应手段是投入十年精力，将假说发展为可被更广泛接受的、科学而严谨的理论。其成果便是盖娅理论。

1983 年，拉伍洛克和他当时的研究生安德鲁·沃森共同开发了名为“雏菊世界”的实验计算机模型[10]——一颗虚构的类地行星以及一个太阳和一套由黑白两种雏菊组成的简单的生态系统。两种雏菊相互竞争生存空间（生态系统可以大致理解为环境和栖居其中的生物体）。这两种雏菊所覆盖的土地比例会影响行星的温度。随着太阳亮度不断增强，就像我们现实中的太阳一样，行星表面的白色雏菊越来越多。白色雏菊比黑色雏菊反射更多的太阳热量，因此能使行星保持较低温度。[11]

该模型证实了同一环境中的两种相互竞争的植物可调节环境温度，并最终达到最理想的温度。许多科学家都同意雏菊世界模型，尤其是数学家们。其中一位拥护者是数学家彼得·桑德斯*，他认为该模型值得研究。另一位拥护者，英国地球系统科学家蒂莫西·伦顿*，发表了若干有关盖娅理论含义及其数学基础的科学论文。

在发展盖娅理论之外，拉伍洛克还撰写了好几部关于盖娅的作品，包括《盖娅时代：地球传记》（1988）[12]。在这本书里，拉伍洛克使用科学用语解释了自己的思想，并回应了有关其第一部作品的批评和评论。在这之后，召开了大量科学会议并讨论盖娅理论。例如 1994 年，来自许多不同学科的学者参与了“可自我调节的地球”的主题会议，共同讨论了盖娅理论以及地球调节过程的方方面面。[13]

冲突与共识

在过去的 40 多年中，拉伍洛克的盖娅理论发展得愈发繁复，也开放、接受修正。从生命为自身利益控制行星状况的启发性假说，到生命是塑造地球的关键因素这一更为强有力的复杂理论，盖娅思想不断进行演化。[14]

地球大气调节的控制论反馈机制概念与全球变暖的辩论相关联，这也是 2001 年《阿姆斯特丹全球变化宣言》所强调的。盖娅理论的环境维度主要围绕两大主题：生物圈遭人为破坏的后果，以及栖息地遭破坏与地球生态系统碎裂化的结果。尽管生物圈可自行吸收并消化小规模干扰，但大规模破坏迟早会引发全球气候深远而不可控的后果。[15]

盖娅思想也受到社会学家们（研究社会及社会行为）的重视。社会学家艾琳·克里斯特 * 和生态学家 H. 布鲁斯·林克 * 写道："当前，人为放大温室效应，其速度之快、规模之大足以引发正反馈……反过来，正反馈可能会导致热度失控。这种可能性不仅会招致人类普遍承受苦难，地球也会沦为一片荒原。"[16]

在这层意义上，盖娅假说可视为"游戏规则变革者"，因为它从根本上改变了人们对地球如何增强可持续发展的看法。

尽管地球是可自我调节的、有生命的整体这一观点尚未达成共识，拉伍洛克的盖娅思想及其核心，尤其是微生物对调节地球大气的作用，已得到科学研究的广泛讨论。拉伍洛克在其自传中写道："如果他们（科学家们）不愿使用盖娅作为其新科学领域的名称，那么我希望他们可以使用'地球系统科学'作为明智之选。"[17]

1. 詹姆斯・E. 拉伍洛克:《消失的盖娅：最终警告》，伦敦：企鹅出版集团，2010年，第111—112页。
2. W. 福特・杜利特尔：“大自然真的是慈母吗？”，载《共同进化季刊》1981年第29期，第60页。
3. 参见理查德・道金斯:《延伸表现型：作为物竞天择最小单位的遗传因子》，牛津：牛津大学出版社，1982年，第1—307页。
4. 参见道金斯:《延伸表现型》，第234—236页。
5. 史蒂芬・杰・古尔德：“克鲁泡特金不是疯子”，载《博物学》1997年第106期，第12—21页。
6. 詹姆士・W. 基什内尔：“盖娅假说：猜想与反驳”，载《气候变化》2003年第58期，第391页。
7. 詹姆士・W. 基什内尔：“盖娅假说：可否测试？”，载《地球物理学评论》1989年第27卷第2期，第223页。
8. 参见詹姆斯・E. 拉伍洛克：“前言”，《盖娅：地球生命的新视野》，詹姆斯・E. 拉伍洛克修订，牛津：牛津大学出版社，2000年，第x—xi页。
9. 艾琳・克里斯特和H. 布鲁斯・林克：“一个宏大的有机整体”，《处于动荡的盖娅：危机时代中的气候变化、生物耗竭与地球伦理》，坎布里奇：麻省理工学院出版社，2010年，第7页。
10. 安德鲁・詹・沃森和詹姆斯・E. 拉伍洛克：“全球环境的生物平衡：雏菊世界的寓言”，载《地球》1983年第35B期，第286—289页。
11. 托比・蒂勒尔:《论盖娅：生命与地球关系的批判性调查》，普林斯顿：普林斯顿大学出版社，2013年，第25页。
12. 詹姆斯・E. 拉伍洛克:《盖娅时代：地球传记》(修订版)，牛津：牛津大学出版社，1995年。
13. 詹姆斯・E. 拉伍洛克:《致敬盖娅：独立科学家的生命》(修订版)，伦敦：纪念品出版社，2014年，第274—275页。
14. 艾琳・克里斯特和H. 布鲁斯・林克：“一个宏大的有机整体”，《处于动荡的盖娅：危机时代中的气候变化、生物耗竭与地球伦理》，艾琳・克里斯特与H. 布鲁斯・林克修订，坎布里奇：麻省理工学院出版社，2010年，第8页。
15. 克里斯特和林克：“一个宏大的有机整体”，第11—12页。
16. 克里斯特和林克：“一个宏大的有机整体”，第14页。
17. 拉伍洛克:《致敬盖娅》，第278—279页。

10 后续争议

要点

- 盖娅假说对气候科学领域贡献卓著，尤其是深度理解全球气候及微生物对调节地球大气环境的作用。
- 多年来，可自我调节的地球这一概念不断启发着科学家、研究人员、学生和政治家，逐渐影响了当代科学。
- 盖娅假说对学术研究和政治的影响广泛，包括地球系统科学这门全新学科的诞生，以及当下辩论的有关全球变暖和人为气候变化的相关性。

应用与问题

在 1979 年发表《盖娅：地球生命的新视野》之后，詹姆斯·E. 拉伍洛克从未中断盖娅的写作。《消失的盖娅》（2010）记录了从 20 世纪 80 年代初到 90 年代中期，要在主流期刊上发表任何相关主题的论文是多么不可能的事。[1] 但在那时之后的许多年里，拉伍洛克具有争议性的假说一直影响着学者、科学家、政治家和公众。在 2013 年的一本关于盖娅假说的批判性作品中，英国地球系统科学家托比·蒂勒尔 * 写道："自问世以来约三十年里，盖娅假说鼓励、激怒了整整一代环境科学家，也激起了他们的好奇心。"[2] 从 60 年代中期至 80 年代末，拉伍洛克的假说受到地质学家、进化生物学家和行星科学家 * 的严厉批评。拉伍洛克采纳了评论家的意见，把他的假说转变为可进行科学测试的、更为复杂的盖娅理论。新修订版的《盖娅》这样记录："在测试之前，盖娅假说只是模糊的推测，离更能被科学接受的盖娅理论还很远。为此，我很感激评论家们。”[3]

拉伍洛克的第一部作品《盖娅》不能说服很多科学家——但有几位学者欣然接受了书中呈现的思想。其中一位是法国裔美国微生物学家勒内·迪博*，他在一篇评论文章中记录了自己读这本书时“极度欣喜”[4]，他认为没有生命的地球将使大气层充满二氧化碳气体，不能维持生命。

1988 年，拉伍洛克发表了其第二部作品《盖娅时代》，相较第一部作品而言更为专业。《盖娅时代》系统地展示其主要思想，例如解释了海藻（一种类似植物的生物体）对气候调节的关键作用。1987 年，拉伍洛克与美国气候科学家罗伯特·查尔森*领衔的三位合作者一起，发表了有关他们称之为克劳假说*的论文。[5] 假说的名字来源于合作者的姓氏首字母（查尔森、拉伍洛克、安德烈埃*与沃伦。假说完善了盖娅概念中的控制论反馈机制，将它归结为浮游植物*这种微小的海洋生物。在“科学指引”数据库中搜索短语“克劳假说”，会得到不同作者提到该主题的 33 篇论文和 16 部著作。

> “盖娅理论的目标是与进化生物学保持一致，认为生物体的进化与其物质环境紧密相连，从而形成单一的、不可分割的过程。”
>
> —— 蒂莫西·M. 伦顿：“盖娅与自然选择”，《自然》

思想流派

20 世纪 80 年代，拉伍洛克与英国海洋与大气科学家安德鲁·沃森共同构建了他们称为“雏菊世界”的数字模型，试图通过可验证的证据将假设转化为科学可靠的理论。

托比・蒂勒尔写道："拉伍洛克年逾九旬，不再那么活跃，其他人则会继续举起前行的火炬。"[6]例如，英格兰埃克塞特大学的地球系统科学家蒂莫西・伦顿自1997年起陆续发表了超过20篇有关盖娅理论的论文。伦顿是沃森的学生。2011年，两人合作出版《创造地球的革命》一书，将盖娅假说的思想继续扩展。[7]伦顿早前还为一篇《自然》期刊1998年收录的有关盖娅理论的评论文章[8]做出重大贡献，他认为自然选择的过程——因查尔斯・达尔文的描述而闻名于世——是组成盖娅不可或缺的一部分。

盖娅模型认为，我们必须把生物体和其所在环境视为一个整体，才能充分理解个性特征（代代相传的特点）持续存在并占主导地位。全世界的科学家和研究者在2001年的《阿姆斯特丹全球变化宣言》中承认："地球系统如同单一的、可自我调节的系统，由物理、化学、生物和人类的成份组成"[9]——即拉伍洛克作品的中心思想，尽管《宣言》并未使用"盖娅"一词。

当代研究

今天，许多学科的科学家都承认盖娅理论是当代科学界最具影响力的理论之一。例如，盖娅理论扩充了达尔文的思想，提出物种的成功取决于生物体的进化与其所在物质环境进化之间的紧密相连[10]。尽管地球科学家詹姆士・基什内尔始终对该假说持批评态度，然而他把它形容为"多产的假说生成器"——为其他研究提供灵感——因为假说"促进了许多有趣的猜想，思考有关生物过程如何在全球范围内促进大气化学和气候的调节。"[11]

今天，拉伍洛克盖娅假说的支持者来自各种不同学科，如天体生物学（探究地球之外的生命）、生物学、地球系统科学、生态学、环境科学和气候科学。安德鲁·沃森（“雏菊世界”模型的共同开发者）和他的学生一起开发了其他数学模型，重现地质年代——也就是在漫长的时期内——大气成分的调节过程。[12]

---

1. 詹姆斯·E. 拉伍洛克：《消失的盖娅：最终警告》，伦敦：企鹅出版集团，2010年，第111页。
2. 托比·蒂勒尔：《论盖娅：生命与地球关系的批判性调查》，普林斯顿：普林斯顿大学出版社，2013年，第ix页。
3. 参见詹姆斯·E. 拉伍洛克：“前言”，《盖娅：地球生命的新视野》，詹姆斯·E. 拉伍洛克修订，牛津：牛津大学出版社，2000年，第xv页。
4. 勒内·迪博：“盖娅与创造性进化”，载《自然》1979年第282期，第154—155页。
5. 格雷格·艾尔斯和吉尔·凯尼：“克劳假说：主要进展回顾”，载《环境化学》2007年第4期，第366—374页。
6. 蒂勒尔：《论盖娅》，第3页。
7. 蒂莫西·迈·伦顿和安德鲁·沃森：《创造地球的革命》，牛津：牛津大学出版社，2013年。
8. 蒂莫西·迈·伦顿：“盖娅与自然选择”，载《自然》1998年第394期，第447页。
9. 詹姆斯·E. 拉伍洛克：“有生命的地球”，载《自然》2003年第426期，第769—770页。
10. 詹姆斯·E. 拉伍洛克：“地球生理学，盖娅的科学”，载《地球物理学评论》1989年第27期，第222页。
11. 詹姆士·W. 基什内尔：“盖娅假说：猜想与反驳”，载《气候变化》2003年第58期，第21页。
12. 詹姆斯·E. 拉伍洛克：《消失的盖娅：最终警告》，伦敦：企鹅出版集团，2010年，第105—122页。

11 当代印迹

要点

- 拉伍洛克的盖娅假说在作品出版近 40 年之后，仍然是科学辩论和发现的热议话题。
- 盖娅理论的一些评论家认为，盖娅理论声称生命体对其所在环境具有控制性，这一思想与查尔斯·达尔文的研究所衍生的进化论不相符合，后者认为生命体应适应环境。
- 其他评论家质疑盖娅理论的预测能力，认为生命实际上可能会摧毁而非拯救这个星球。

地位

如同《盖娅：地球生命的新视野》所阐述的，詹姆斯·E. 拉伍洛克激进的盖娅假说对当代科学家和学者而言一直是具有煽动性的科学话题。即使 40 年后的今天，来自不同领域的科学家仍然在不断重新审视并重新解释该假说。[1]

盖娅理论可跨学科范围广，涵盖从天体生物学到生态学等各种学科。在 30 年里，该理论对多项科学研究以及天体物理学、生物学、地球科学和生态学发现都做出重要贡献。盖娅理论背后的概念引发了地球生理学这门新兴的多学科科目的发展，它又被称为地球系统科学，视地球为相互关联的系统，以期更好地理解物理、化学、生物和人类之间的相互作用对行星过去、现在和未来状态的决定意义。

盖娅理论对于气象科学 * 和环境科学 * 的另一值得注意的贡献，在于其帮助了我们更好地理解全球变暖和气候变化的当代争论。生态哲学，又称深层生态学*，即一种看待接合思想、感受、精神性和行动这些世界性问题的方式，这门学科便深受盖娅假说影响。同样，生态精神*的思想，将生态科学与精神性相联系，也受到拉伍洛克作品的启发。

盖娅假说的当代评论家南安普顿大学地球系统科学教授托比·蒂勒尔认为，盖娅假说达到了一定的科学地位。[2] 然而，虽然一些人已欣然接受该假说，但它仍然会引发激烈的辩论。[3] 另一位该假说的执着评论家、地球科学家詹姆斯·基什内尔在科学期刊上发表研究论文，[4、5] 反驳该理论的预测能力。盖娅假说在解释地球上在全球范围内运作的若干核心机制（例如大气气体的调节）上具有重要性，这保证了其在现代科学中的持续相关性。[6]

> “盖娅理论提出，生物体不断破坏其所在环境，当反馈掉转头来时，将最终导致严峻的后果。我们当前经历的反馈形式有气候变化、臭氧消耗、内分泌干扰和沙漠化。”
>
> ——艾琳·克里斯特和H. 布鲁斯·林克：“一个宏大的有机整体”，《处于动荡的盖娅》

互动

例如地质学、进化生物学和行星科学等学科的科学家对可自我调节、有生命的地球这一思想持严肃保留意见。主要的质疑在于，盖娅假说认为生物控制着地球环境，也即环境受生物区——环境中的动植物——控制。这与进化论相违背。进化论认为，生物体适应其所在环境。[7] 查尔斯·达尔文的自然选择进化论认为，带有有利于生存的特征的生物体易于繁殖，并将那些特性传递给后代，无法生存并繁殖的生物体则会走向灭绝。[8]

蒂勒尔和其他学者认为自然选择的运作规则很简单，时空内效果最好的就会受到青睐，而不考虑未来对更广阔的生态环境可能造成的结果或全球范围的影响。[9] 此外，全球生物区并非密切相连的小家庭，所以无法以整个地球为规模开展合作。[10]

持续争议

在 2013 年作品《论盖娅：生命与地球关系的批判性调查》中，蒂勒尔仔细考虑了盖娅假说为何对一些科学家及公众能产生经久不衰的吸引力。首先，他认为盖娅是具有全局观的一门科学，可以为例如为何地球能在很长的时间里一直保持适宜居住这样深层次的问题提供解答。[11]

其次，盖娅提出了我们的星球在未来如何应对全球变暖的机制。詹姆士·基什内尔和托比·蒂勒尔都认为，为了保护作为生命保障系统的地球，我们对于其自然过程的理解必须基于正确的观点。在这一背景下，基什内尔提供了一些例子来质疑盖娅理论的预测能力。例如，盖娅理论预测生物过程严格地调节了地球大气的组成，但相比工业化之前，微生物的碳吸收率仅增加了 2%，而大气中的二氧化碳气体却增加了 35%。[12] 拉伍洛克认为，如今地球的自我调节系统已远不足以弥补人为的 * 温室气体污染。

盖娅假说对当代学术界影响的另一实例便是彼得·沃德 * 称之为“美狄亚假说” *[13] 的反盖娅概念。他是美国古生物学家 *，专门研究化石，任职于澳大利亚的阿德莱德大学。与盖娅假说一样，美狄亚假说以古希腊神话中的人物美狄亚命名（美狄亚是神话英雄伊阿宋的妻子）。沃德认为生命具有自我毁坏性，还举出了地球历史

上的大灭绝 * 作为例子。沃德的论点——本质上是盖娅假说的对立面——是生命使生物圈的温度不断上升，并在约十亿年之后导致其自身的终结。[14]

---

1. 托比·蒂勒尔：《论盖娅：生命与地球关系的批判性调查》，普林斯顿：普林斯顿大学出版社，2013 年，第 1—6 页。
2. 蒂勒尔：《论盖娅》，第 1—6 页。
3. 蒂勒尔：《论盖娅》，第 3 页。
4. 詹姆士·W. 基什内尔："盖娅假说：可否测试？"，载《地球物理学评论》1989 年第 27 卷第 2 期，第 223 页。
5. 詹姆士·W. 基什内尔："盖娅假说：猜想与反驳"，载《气候变化》2003 年第 58 期，第 21—45 页。
6. 克里斯平·蒂克尔："科学家论盖娅"，载《金融时报》2002 年，登录日期 2013 年 12 月 23 日，http://www.crispintickell.com/page19.html。
7. 蒂莫西·迈·伦顿："盖娅与自然选择"，载《自然》1998 年第 394 卷，第 439—447 页。
8. 查尔斯·达尔文：《论依据自然选择即在生存斗争中保存优良族的物种起源》，伦敦：约翰·默里出版公司，1859 年。
9. 蒂勒尔：《论盖娅》，第 34 页。
10. 蒂勒尔：《论盖娅》，第 40 页。
11. 蒂勒尔：《论盖娅》，第 1—6 页。
12. 基什内尔："盖娅假说：猜想与反驳"，第 21 页。
13. 彼得·沃德：《美狄亚假说：地球上的生命会最终自我毁灭吗？》，普林斯顿：普林斯顿大学出版社，2009 年，第 208 页。
14. 莫伊塞斯·维拉斯克斯·马诺夫："美狄亚假说：回应盖娅假说"，载《基督教科学箴言报》2010 年 2 月 12 日，登录日期 2013 年 12 月 28 日，http://www.csmonitor.com/Environment/Bright-Green/2010/0212/The-Medea-Hypothesis-A-response-to-the-Gaia-hypothesis。

12 未来展望

要点

- 盖娅假说在过去的四十年间影响了科学理论和发现，并与全球气候变化为主题的研究息息相关。
- 今天，气候科学家在建立地球未来气候的模型时，都会考虑盖娅假说的建议，即拉伍洛克开创性工作的重要影响。
- 《盖娅：地球生命的新视野》中给出的思想带来了地球系统科学这门全新学科的发展。

潜力

由于发表了《盖娅：地球生命的新视野》等作品，詹姆斯・E. 拉伍洛克被美国流行杂志《滚石》评为20世纪最具影响力的科学家之一。[1] 该作品受众为非科学领域的读者，根据其文本所述，生物体及其所在的物理环境构成了一个完整的实体，控制着地球大气和气候。可自我调节的地球这一思想在当时相当激进，许多科学家抨击盖娅假说是伪科学。1988年，拉伍洛克出版了其第二部作品《盖娅时代》，专为科学家而作。

随着盖娅理论开始不仅受到公众而且更受到政治家、学者与科学家的关注，拉伍洛克还就盖娅假说创作了五本书[2]。英国《每日电讯报》的一篇文章认为这些作品“是20世纪最重要的论战之一”[3]（此处的“论战”是指以说服为目的的措辞强硬的文章）。

“今天，盖娅假说越来越受到广泛赞誉，”托比・蒂勒尔在2013年出版的评论作品《论盖娅》中这样写道。[4] 盖娅假说有潜力去影响关于地球气候及其如何进行调节的辩论。在《盖娅》及后续作品中，拉伍洛克一直在强调人为的（人类导致的）全球变暖。工业化和石油、天然气与煤炭的燃烧增加，导致了温室气体（尤其是二氧化碳）浓度增加。随着温室效应增强，大气留住了更多热量，而以冰和雪为主的反光表面的减少，意味着吸收更多光线（和热量），而反射的光线更少，可能导致形成正反馈效应并引发受热不均匀。[5]

在这些情况下，盖娅假说为科学家和决策者提供了看待现存环境问题的独特视角。盖娅假说的确对理解全球变暖问题作出了重要贡献。[6]

> "我们需要全世界的人一同感受目前的真实危险，这样大家才会自发动员并不遗余力地进行有序而可持续的退让，从而达到人类与盖娅和谐共存。"
>
> —— 詹姆斯 · E. 拉伍洛克：《盖娅的复仇》

未来方向

拉伍洛克在《消失的盖娅：最终警告》（2010）中写道，地球气候变化将导致敏感的生态系统消失不见，并有可能威胁人类的生命。[7] 拉伍洛克严厉批评了一些气候科学家和政治家，因为他们没有把地球看作可自我调节的、有生命的实体。[8] 在很长一段时间里，他们把地球看作一块坚石，而没有考虑生命（生物的）与地球物理环境之间的相互作用。[9] 最近，政府间气候变化专门委员会已着手思考新的气候模型，即地球系统模型，确认了影响气候的多种因素之间的相互影响。

根据英国《卫报》的一篇文章，拉伍洛克“并非散布恐怖威胁论之人，而是解决问题的务实之人，在多种层面提供了缓解气候危机的建议。”[10] 拉伍洛克深信盖娅的理念将有更进一步的发展。例如，地球系统科学家蒂莫西·伦顿就盖娅假说开展研究。在其最近出版的《消失的盖娅》一书中，拉伍洛克声称例如地质工程*——对地球自然系统的大规模故意干预——这样的新技术将会出现，以抵制气候变化。

小结

自 1979 年《盖娅：地球生命的新视野》出版以来，不少盖娅假说的应用都一一得到实现。拉伍洛克在书中为进一步开展学术研究提供了大量有趣的想法。其盖娅理论又促进了一些有趣的阐释，阐释生物过程如何调节全球规模的大气化学成分与气候。[11] 在《盖娅》一书中，拉伍洛克认为地球是一个可自我调节的系统。例如大气的气体、大气温度和海洋的盐浓度的调节这样一些行星机制都在书中做了解释。拉伍洛克还说明了这些系统如何通过控制论反馈机制——根据变化进行调整的自动控制系统——从而受到地球生物体的控制。拉伍洛克在《盖娅》中记录的这些思想，对过去超过 40 年间的生态学及气候科学领域学术研究与应用作出了重要贡献。

盖娅假说的本质是跨学科的，吸收了多门不同学科的目标和知识。它无法被归入某一单独的学科。拉伍洛克建议应在地质生理学或地球系统科学这门新兴学科下研究地球系统。[12] 例如生态学、海洋生物学和气候科学等其他学科的学者也在思考盖娅的思想。对盖娅假说最非凡的应用之一是开发了名为“雏菊世界”的数学模型[13]——生态学家现在用该模型来测试生物多样性与生态系统稳定性对健康的生存环境的作用。

最新的盖娅假说应用是在学校教育中。[14]2009年在巴西的一项研究表明，盖娅假说对学校科学教育中理解人类活动和全球变暖及气候变化等当代环境问题作出了贡献。研究还发现，拉伍洛克具有争议性的盖娅假说的跨学科性质，是学校跨学科学习有趣且有效的工具。

---

1. 杰夫·古德尔："预言家詹姆斯·拉伍洛克"，载《滚石》2007年11月1日，登录日期2013年12月23日，http://www.rollingstone.com/politics/news/james-lovelock-the-prophet-20071101。
2. 詹姆斯·E.拉伍洛克：《致敬盖娅：独立科学家的生命》(2000年)，《为盖娅疗伤：行星地球的实用医学》(2001年)，《盖娅：病态行星的一剂良药》(2005年)，《盖娅的复仇：地球为何在还击——我们怎样才能拯救人类》(2007年)和《消失的盖娅：最终警告》(2010年)。
3. 詹姆士·弗林特："地球——最终之战"，载《每日电讯报》2006年2月6日，登录日期2013年12月23日，http://www.telegraph.co.uk/culture/books/3649909/Earth-the-final-conflict.html。
4. 托比·蒂勒尔：《论盖娅：生命与地球关系的批判性调查》，普林斯顿：普林斯顿大学出版社，2013年，第ix页。
5. 艾琳·克里斯特和H.布鲁斯·林克："一个宏大的有机整体"，《处于动荡的盖娅：危机时代中的气候变化、生物耗竭与地球伦理》，艾琳·克里斯特和H.布鲁斯·林克修订，坎布里奇：麻省理工学院出版社，2010年，第1—20页。
6. 詹姆斯·E.拉伍洛克：《通往未来的艰难之旅》，伦敦：企鹅出版集团，2015年，第85—103页。
7. 詹姆斯·E.拉伍洛克：《消失的盖娅：最终警告》，伦敦：企鹅出版集团，2010年，第1—45页。

8. 詹姆斯·E. 拉伍洛克:《盖娅的复仇:地球为何在还击——我们怎样才能拯救人类》,伦敦:企鹅出版集团,2007 年,第 61—83 页。
9. 拉伍洛克:《消失的盖娅》,第 1—45 页。
10. 彼得·福布斯:"吉姆会处理",载《卫报》2009 年 2 月 21 日,登录日期 2013 年 12 月 23 日,http://www.theguardian.com/culture/2009/feb/21/james-lovelock-gaia-book-review。
11. 詹姆斯·W. 基什内尔:"盖娅假说:猜想与反驳",载《气候变化》2003 年第 58 期,第 21 页。
12. 詹姆斯·E. 拉伍洛克:"地球生理学,盖娅的科学",载《地球物理学评论》1989 年第 27 期,第 215—222 页。
13. 安德鲁·詹·沃森和詹姆斯·E. 拉伍洛克:"全球环境的生物平衡:雏菊世界的寓言",载《地球》1983 年第 35B 期,第 286—289 页。
14. 里卡多·桑托斯·多卡尔莫、内伊·弗雷塔斯·努内斯·内托和沙尔贝勒·尼诺·哈尼:"巴西高中生物学课本中的盖娅理论",载《科学与教育》2009 年第 18 期,第 469—501 页。

术语表

1. **藻类**：通过光合作用吸收阳光作为养分的类植物生物。

2. **利他主义**：显示想要无私帮助他人的人类行为。

3. **美国地球物理学联合会**：致力于地球物理学研究进步与更广泛理解的非营利组织。

4. **氨**：一种强烈而无色的气体，由氮和氢元素组成，具有特殊的刺鼻气味。

5. **《阿姆斯特丹全球变化宣言》**：来自四个国际全球变化研究项目的科学界人士发表的声明，承认人类对全球环境不断增多的改变，加上气候变化的威胁，对人类福祉会造成巨大影响。

6. **人为的**：人类导致或受人类影响的。

7. **阿波罗任务**：美国国家航空航天局（NASA）实行的航天计划。阿波罗宇航员成为了1969年至1972年间最早登陆月球的人。

8. **人造卫星**：绕地球和其他太阳系行星飞行的人造物体。

9. **天体生物学**：研究宇宙中地球之外生命起源与进化的科学。

10. **天文学家**：研究恒星、行星、卫星、彗星、星系等的科学家。

11. **天体物理学**：天文学分支，研究恒星、卫星、行星等天体的物理特性和组成。

12. **大气气体**：地球大气中存在的气体，如氧气、二氧化碳、氮气等。

13. **细菌**：人体内外到处可寻的单细胞微生物。某些细菌对人类健康有益，但大多数有害。

14. **生物化学家**：具有生物化学——研究生物体化学过程的学科——资质的科学家。

15. **生物多样性**：某一地区生物的多种多样（多样化）。

16. **生物学**：动植物等生物的科学研究。

17. **生物学家**：关注包括动植物在内的生物体的科学家。

18. **生物圈**：地球或其他被生物体占据的行星的地表和大气环境区域。

19. **生物区**：一个地区、栖息地或地质年代的动植物等生物体。

20. **二氧化碳**：地球大气中一种常见的无色无味不可燃气体。它是在呼吸过程中以及有机物分解与燃烧过程中产生的一种温室气体。

21. CFCs：参见以下“氯氟烃”。

22. **化学**：研究物质化学特性及其反应的学科。

23. **氯氟烃**：又称为 CFCs，无毒、不易燃的化合物，含碳、氯和氟原子，常用作冷却剂。

24. **克劳假说**：拉伍洛克和其他三位科学家阐述的克劳假说，提出海洋表面的微生物可调节其自身种群。其命名由罗伯特·查尔森、詹姆斯·拉伍洛克、迈因拉特·安德烈埃与斯蒂芬·G. 沃伦这四位科学家的姓氏首字母组成。

25. **气候变化**：行星天气模式的长期变化，例如平均温度改变。

26. **气候科学**：研究气候——相当长一段时间内日常天气的平均情况。

27. **大陆漂移**：德国科学家阿尔弗雷德·魏格纳提出的理论，认为地球表面在数百万年间不断进行漂移。

28. **宇宙辐射**：从地球大气层之外发散的高能亚原子尺寸粒子所组成的宇宙射线。

29. **控制论反馈机制**：机械、物理、生物或社会系统中的反应，影响该系统的持续活跃性或生产率。

30. **雏菊世界**：拉伍洛克及其学生安德鲁·沃森设计了一个数字模型，为盖娅假说提供可测试的证明。在模型中，雏菊世界是假想的一颗晴朗无云的类地行星，大气层的温室效应几乎可以忽略不计。行星上有两种雏菊——一种黑色，一种白色。黑色（深色）雏菊覆盖的土地所反射的光线比空地要少；白色（浅色）雏菊覆盖的土地所反射的光线比空地要多。

31. **深层生态学**：包含感觉、精神、思想和解决世界问题的行动的一切方式。它涉及超越现代文化以自我为中心的本质，而把人类视为地球的组成部分。

32. **非稳定**：丧失或缺少均衡或稳定性，在这种情况下，对立的力量或影响处于不平衡状态。

33. **地球科学**（Earth science）：利用地质学、化学、生物与气候学等多种领域的学科，来研究我们所在行星的深厚历史和系统性运作。

34. **地球系统模型**：观察大气、海洋、陆地、冰层与生物圈相互作用的模型，以测量不同情况下地区及全球气候条件的状态。

35. **地球系统科学**：把地球视为一体化系统的跨学科研究，寻找一种方式来理解（物理、化学和生物）环境与人类活动之间的相互作用，建构行星在过去、现在和未来的状态。

36. **地球系统科学家**：地球系统科学领域的科学家。

37. **生态学**：研究生物体之间及其与所在物理环境关系的生物学分支。

38. **生态精神**：基于信仰自然、地球和宇宙神圣性的一种哲学。

39. **生态系统**：某一特定物理环境中发现的生物体所组成的生物系统，生物体之间及其与所在环境之间相互产生作用。

40. **电子捕获检测器**：检测样本中微量化合物的仪器。

41. **环境科学**：观察环境条件及其对生存其中的生物体影响的跨学科领域。

42. **平衡**：对立的力量或影响达到相互平衡的状态。

43. **蒸发**：由于温度或压力上升，或两者同时上升，物质从液体变为气体的过程。

44. **蒸发岩**：咸水蒸发形成的矿物沉淀。

45. **进化**：一类动物或植物的特性一代接一代地逐渐发生的变化，它解释了现存物种从不同祖先变化而来的过程。

46. **进化生物学**：研究生物体进化尤其是分子进化及微生物进化的生物学分支。其他研究领域还包括行为、遗传学、生态学、生活史和变化发展。

47. **盖娅假说**：认为生命存在本身主动使地球表面、大气和海洋的物理与化学状况变得适宜生存的思想。根据该思想，正如詹姆斯·拉伍洛克所说，“包括生命在内的整个地球表面是一个可自我调节的实体。”

48. **盖娅理论**：盖娅假说衍生而来的理论，詹姆斯·拉伍洛克与安德鲁·沃森开发的数学模型（“雏菊世界”）可对其充分进行证明。

49. **盖娅思维**：认为包括人类在内的全部地球生物相互作用、塑造其生存环境并使环境适宜生存的哲学立场——思维方式整体受盖娅假说启发。

50. **地球化学家**：研究地球或其他行星的固体物质的化学成分与化学变化的学者。

51. **地质工程**：对地球自然系统采取的故意的、大规模的技术干预。地质工程通常被认为是应对全球气候变化的技术手段。

52. **地质学**：地球起源、历史和结构的研究——涉及固态地球及其岩石的科学。

53. **地球物理学家**：研究定义行星的各种引力、磁性、电气和地震现象（如地震）的学者。

54. **地球生理学**：对地球全部生物体之间相互作用的研究。地球生理学是研究全球规模的问题的跨学科场所。

55. **地球科学**（Geoscience）：地质学、地球物理学和地球化学等研究地球的科学。

56. **全球变暖**：地球大气长期平均温度的逐步上升。

57. **目标导引**：通过对给定信息进行各种“如果——那么”的假设性分析从而得到计算结果的过程。

58. **希腊神话**：古希腊的神话和传说。希腊神话主要涉及神祇和英雄、各种其他神兽以及古希腊崇拜和仪式的起源与重要性。

59. **温室效应**：地球大气层吸收太阳能量，从而使大气层足够温暖并维持生命的自然过程。之所以称之为温室效应，是因为如同温室一般将热量留在自身内部，工作原理与之基本相同。

60. **温室气体**：吸收肉眼不可见的红外线辐射从而导致温室效应的气体。

61. **动态平衡或动态平衡状态**：生物所栖居环境不断改变但其本身保持稳定的状态。

62. **人文主义**：强调人文关怀的哲学立场，其激进形态是源自宗教的一种伦理制度。

63. **假说**：尚未得到科学证实，但会引发进一步研究或讨论的思想或概念。在科学领域，假说需经过大量测试之后才能成为理论。在科学领域之外，假说和理论两者通常可互换使用。

64. **《伊卡洛斯》**：行星科学领域的科学期刊。

65. **政府间气候变化专门委员会（IPCC）**：在联合国支持下各国代表组成的政府间科学团体。

66. **海洋生物学**：对海洋及其他咸水环境中生存的海洋物种或生物体的研究。

67. **大灭绝**：全球规模的动植物多样性相对突然的减少。在地球生命的历史上发生过几次大灭绝，其发生的地质时期通常较短。

68. **美狄亚假说**：认为生命是自身的敌人且几乎所有的地球大灭绝都是

由生命自身导致的思想。美国古生物学家彼得·沃德提出了该假说，与拉伍洛克的盖娅假说完全对立，后者认为生命自身使地球维持着适宜生存的状态。盖娅视地球为“良母”，而美狄亚视地球为“恶母”。

69. **隐喻**：通常用于比较两个不同对象、思想、思维或感受的词或短语。

70. **甲烷**：一种无色无味的可燃气体。甲烷是天然气的主要成分，也是最简单的碳氢化合物。植物或其他有机物分解时会释放甲烷。甲烷还是一种温室气体。

71. **美国国家航空航天局（NASA）**：负责民用太空计划的美国政府机构。

72. **国家医学研究所**：位于英国伦敦的一家医学研究机构。由英国医学研究委员会于1913年建立。

73. **自然选择**：动植物等具有某种特性的单个生物体在同一种群中比其他个体具有更高生存率或繁殖率的进化变化过程。较成功的个体将把那些有用的可遗传特点传递给其后代子孙。

74. **生物体**：动物、植物、菌类或细菌等有生命的生物实体。

75. **古生物学**：研究动植物化石的科学研究。

76. **光合作用**：绿色植物和其他生物体利用光能（通常来自阳光）通过二氧化碳和水创造养分的过程。植物用这种方式生产食物，氧气作为废品得以释放。

77. **物理**：研究物质和能量性质和特性的基础科学分支。

78. **浮游植物**：微小的海洋植物，在平衡的生态系统中为水母、鲸和蜗牛等在内的各种海洋生物提供食物。

79. **行星科学**：地球等行星、卫星和行星系统，尤其是太阳系内行星、卫星和行星系统的研究。

80. **行星科学家**：研究行星的学者。

81. **辐射**：以波或粒子流形式传输的能量——例如火的热和光就是一种

辐射形式。

82. **放射性**：原子衰变产生的辐射或能量波的散发。

83. **简化论**：将复杂的事物看作由几部分拼成的总和，而不考虑组成部分之间可能的相互影响。

84. **科学指引**：提供（自然）科学和医学领域由同行评议的期刊论文和图书章节的科学数据库。

85. **可自我调节的地球**：认为所有的生物体与地球空气、水和岩石进行相互作用并使地球以稳定趋势适宜生存的观点。

86. **苏联**：又称苏维埃社会主义共和国联盟（USSR），位于欧洲大陆的一个社会主义国家。苏联存在时间为 1922 年至 1991 年，是由共产党领导的国家。其首都为莫斯科。

87. **太空探索**：研究地球范围之外的生物物理环境——太空、恒星、行星、彗星等——主要通过使用载人宇宙飞船、人造卫星或探测器。

88. **Sputnik（人造卫星名）**：1957 年发射至太空的俄罗斯卫星。它是首个离开地球大气层的人造物体。

89. **超个体**：由许多独特生物组成的实体。

90. **目的论的**：有关目的论——解释理由的方法，或依照目的、目标作解释。

91. **神学**：神祗和宗教信仰的研究。

92. **理论**：一组事实或现象的自然解释。理论通常具有连贯性、预言性、系统性，且应用广泛。理论一般可经受检验，随着得出更多信息可不断得到改进或修正，因此理论的预测能力随着时间推移而变得更为强大。

93. **恒温器**：如烤箱等电器设备中自动调节温度的装置。

94. **紫外线**：人眼无法看到的部分光谱。

95. **联合国**：1945 年建立的国际机构，以促进国际合作、和平和安全为目标。

96. **维京火星探测计划**：美国国家航空航天局的火星探索任务。探测器于 1975 年发射，是第一架在火星表面安全着陆并送回图像的探测器。

97. **沃拉斯顿奖**：伦敦地质学会颁发的地球科学领域最高奖项。获奖的地质学家均在地球科学的基础和 / 或应用方面做出杰出贡献。

98. **工人阶级**：社会经济学术语，用以形容工作收入较低、技术含量有限且教育要求不高的社会阶层人士。

人名表

1. **迈因拉特·安德烈埃**（1949年生），德国矿物学家，现任德国马克斯·普朗克化学研究所生物地球化学系主任。

2. **佩内洛普·J. 波士顿**，新墨西哥矿业及科技学院地球与环境科学系教授。波士顿因提出将小型跳跃机器人送至火星用以促进探索而闻名。

3. **罗伯特·查尔森**，华盛顿大学大气学与化学荣誉教授。查尔森与詹姆斯·拉伍洛克以及其他两位科学家——迈因拉特·安德烈埃和斯蒂芬·G. 沃伦共同合作提出了克劳假说。

4. **艾琳·克里斯特**，社会学家，弗吉尼亚理工学院暨州立大学（弗吉尼亚理工大学）社会科学与技术副教授。

5. **查尔斯·达尔文**（1809—1882），英格兰博物学家、地质学家。达尔文因其自然选择进化论而闻名。

6. **理查德·道金斯**（1941年生），英国进化生物学家、知名作家，直言不讳的无神论者。道金斯曾任牛津大学教授，现为牛津大学新学院退休荣誉教员。

7. **福特·杜利特尔**（1941年生），生物化学家，出生于美国伊利诺伊州。现在是加拿大戴尔豪斯大学教授，长期就拉伍洛克的盖娅假说作评论。

8. **勒内·迪博**（1901—1982），法国裔美国微生物学家，洛克菲勒大学荣誉教授。迪博是盖娅假说主要思想的早期支持者之一。

9. **内莉·A. 伊丽莎白**，詹姆斯·拉伍洛克的母亲，职业为私人秘书。她曾带拉伍洛克到当地图书馆借阅科幻小说。

10. **悉尼·埃普顿**，任职壳牌英国桑顿研究中心，在20世纪70年代与詹姆斯·拉伍洛克合作共同发展盖娅假说。

11. **威廉·戈尔丁**（1911—1993），英国小说家，1983 年获诺贝尔文学奖。他是拉伍洛克所在村庄的居民。他提议以“盖娅”命名拉伍洛克的假说。

12. **史蒂芬·杰·古尔德**（1941—2002），美国古生物学家、进化生物学家，职业生涯大部分时间在哈佛大学授课。古尔德对拉伍洛克的盖娅假说颇多批评。

13. **瓦茨拉夫·哈维尔**（1936—2011），捷克剧作家，1989 年至 1992 年担任前捷克斯洛伐克总统，1993 年至 2003 年担任捷克共和国总统。

14. **戴恩·希契科克**，哲学家，曾与詹姆斯·拉伍洛克一同在美国国家航空航天局工作。他们仔细研究了火星的大气数据，认为火星上没有生命存在。十年之后，火星飞行任务证实了他们的结论是正确的。

15. **海因里希·霍兰**（1927—2012），出生于德国，后定居美国。霍兰是哈佛大学退休荣誉教授，对地球化学的理解作出重要贡献。

16. **乔治·伊·哈钦森**（1903—1991），美国动物学家，以淡水湖的生态研究而知名。哈钦森出生于英格兰，就读于剑桥大学。1928 年，他加入耶鲁大学教授动物学，其大部分职业生涯都在那里度过。

17. **詹姆士·赫顿**（1726—1797），具有重要影响力的苏格兰地质学家，其思想早于拉伍洛克的思想。

18. **托玛斯·H. 赫胥黎**（1825—1895），英格兰生物学家，以其有关查尔斯·达尔文自然选择进化论的作品最为出名。赫胥黎在使达尔文的理论被科学家和公众接受这一方面做出最为杰出的努力。

19. **约翰·肯尼迪**（1917—1963），又称为 JFK，美国第 35 任总统（1961—1963），是入主白宫时最为年轻的总统。1963 年 11 月 22 日，他在德克萨斯州的达拉斯遇刺，成为去世时最年轻的总统。

20. **詹姆士·基什内尔**，现任加利福尼亚大学伯克利分校地球与行星科学教授。基什内尔对盖娅假说一直充满兴趣，同时持批评态度。

21. **叶夫格拉夫·M. 科罗连科**，19 世纪乌克兰哲学家和科学家。科罗连科自学成才，对当时重要的自然科学家的作品颇为熟悉。

22. **蒂莫西·伦顿**，埃克塞特大学气候变化与地球系统科学教授。终其研究生涯，伦顿对詹姆斯·拉伍洛克具有争议性的盖娅假说充满兴趣，被认为可能是拉伍洛克的继承人。

23. **托马斯·A. 拉伍洛克**，詹姆斯·拉伍洛克的父亲。托马斯·拉伍洛克的职业是艺术商，同时对自然保持着强烈的感情。

24. **林恩·马古利斯**（1938—2011），马萨诸塞大学阿默斯特分校杰出的地球科学教授。马古利斯与拉伍洛克合作，共同发展了具有争议性的盖娅假说。

25. **玛丽·米奇利**（1919 年生），英格兰道德哲学家，其特殊兴趣涵盖科学、伦理学、人性和动物权利。米奇力在写作中赞同拉伍洛克盖娅假说的道德解读。

26. **万斯·小山**（1922—1998），美国国家航空航天局从事火星生命探索的生物化学家，他因对阿波罗登月计划所得样本的开创性的生命探测试验而被铭记。

27. **H. 布鲁斯·林克**，生态学家，弗吉尼亚州河谷保护理事会执行理事。

28. **彼得·桑德斯**，伦敦国王学院数学荣誉教授。

29. **斯蒂芬·施奈德**（1945—2010），美国人，斯坦福大学环境生物学与全球变化科学教授。他对全球气候变化主题的研究、政策分析和外展服务受到国际公认。

30. **爱德华·修斯**（1831—1914），奥地利地质学家，对其领域的研究作出杰出贡献。许多概念的诞生都归功于修斯，这些概念导致了板块构造（地壳异动）理论和古地理学（古代地球大陆地块的研究）的产生。

31. **托比·蒂勒尔**，南安普顿大学地球系统科学教授。蒂勒尔以其对盖娅假说的批判性评论而闻名。

32. **弗拉基米尔·伊·维尔纳茨基**（1863—1945），俄国知名矿物学家和地质化学家。维尔纳茨基被认为是现代地球化学的创始人。

33. **儒勒·凡尔纳**（1828—1905），19 世纪法国小说家和诗人。他是《环游世界八十天》和《海底两万里》的作者。

34. **彼得·沃德**（1949 年生），美国古生物学家，现任澳大利亚阿德莱德大学教授。他以 2009 年出版的反盖娅假说《美狄亚假说》闻名于世。

35. **安德鲁·沃森**（1952 年生），英国海洋与大气科学家，现任埃克塞特大学教授。沃森是詹姆斯·拉伍洛克带的博士研究生。他们共同开发了“雏菊世界”这一计算机模型，为证实生物体可调节其所在环境提供了证明。

36. **阿尔弗雷德·魏格纳**（1880—1930），出生于德国柏林，极地研究员、地球物理学家、气象学家。魏格纳因开创性的大陆漂移理论闻名于世，其理论首次提出所有的大陆都在地球上缓慢移动。

37. **赫·乔·威尔斯**（1866—1946），英国小说家。《时间机器》和《世界大战》的作者。威尔斯以其科幻小说最为出名。

WAYS IN TO THE TEXT

KEY POINTS

- James E. Lovelock, one of the world's foremost independent scientists and an outstanding popular science book writer, was born in 1919 into a working-class* family in the southeast of England.
- According to Lovelock's *Gaia: A New Look at Life on Earth*, the planet's living beings interact with the atmosphere, oceans, and rocks to form a self-regulating,* stable biosphere* where life can flourish ("biosphere" here refers to those parts of a planet occupied by living beings).
- Lovelock's controversial Gaia hypothesis*—according to which the Earth is a living superorganism* (an entity, like a colony, made up of many distinct individuals or organisms)—has led to the development of an entirely new academic discipline: Earth system science.*

Who Was *James E. Lovelock?*

James E. Lovelock, the author of *Gaia: A New Look at Life on Earth* (1979), was born in 1919 in the southeast of England into a working-class family.[1] His father Thomas A. Lovelock* was an art dealer; his mother Nellie A. Elizabeth* worked in city administration as a personal secretary;[2] neither had any serious formal education. Nellie had to leave school at the age of 13 to earn her living while Thomas had never been to school as a child, and could not read or write before he attended a technical college later in his life.[3] Perhaps on account of their history, Lovelock's parents valued education highly and always encouraged their son to go to school—but financial hardship in the family meant he could not afford to go to college when he came of age. Instead, he took a training position with a firm of chemical consultants while attending evening classes.

Lovelock eventually went to the University of Manchester[4] with a scholarship and graduated in chemistry* in 1941. In 1948, he received his doctor's degree in medicine. In 1959, he received a doctorate in biophysics (a discipline in which the science of physics is applied to the study of organisms) from the University of London.[5]

Working for most of his life as an independent scientist, Lovelock has been described by the British *Guardian* newspaper as a "world-famous author and speaker."[6] He owes this fame to his radical Gaia hypothesis: the proposition that the Earth can be considered an entity composed of many distinct individuals or organisms—a living superorganism comparable to a colony of bees or ants. Referring to Charles Darwin,* the nineteenth-century founder of evolutionary* theory, the English philosopher Mary Midgley* has written of Lovelock, "Though fanatically accurate over details, he never isolates those details from a wider, more demanding vision of their background. He thinks big. Preferring, as Darwin did, to work outside the tramlines of an institution, he has supported himself since 1963 through inventions and consultancies."[7]

What Does *Gaia: A New Look at Life on Earth* Say?

Having discussed the idea in academic papers from 1969 onward,[8] Lovelock began to write *Gaia: A New Look at Life on Earth* in 1974, while living in an area of natural beauty in western Ireland. The book, his first, is written in simple, nontechnical language to convey his ideas to a nonacademic audience; in it, he introduced the general public to the rather controversial idea of our planet as a self-regulating and living entity, which he called "Gaia" after the ancient Greek* goddess of the Earth.

For Lovelock, humans are not a special feature of the planet, being part of a broader community of living creatures. Referencing the political and philosophical position of humanism,* emphasizing the importance of human affairs, Lovelock writes, "I began more and more to see things through her [Gaia's] eyes and slowly dropped off, like an old coat, my loyalty to the humanist Christian belief in the good of mankind as the only thing that matters."[9]

According to Lovelock's hypothesis,* the living Earth is a superorganism—a living thing made up of many other living things all interacting both mutually and with the air, the oceans, and the planet's surface rocks. This system of mutual interaction has the effect of making Earth a fit and comfortable place to live. Suggesting for the first time that living organisms control their nonliving environment, or surroundings, the idea has proven to be controversial.

Lovelock laid out his thoughts as a story in a series of interlinked chapters. Because of his simple writing style, shortage of evidence, and his use of myth and poetry, the book received very harsh reviews[10] within the scientific community when it was first published in 1979. Scientists from several disciplines, among them biology* and geology,* expressed serious reservations about the idea. (Biology is the study of living organisms; geology is the study of the formation and structure of the planet's physical material, such as rock.)

The essence of Lovelock's Gaia hypothesis is that the entire surface of the Earth, including all living organisms, is a self-regulating body. It makes changes to itself as needed to maintain a balance between its physical, chemical, and biological environments, thereby sustaining life and helping it evolve through time. Earth's lasting tendency to maintain a stable condition for its living creatures through self-regulation is known as "homeostasis."* This homeostatic feature of the planet Earth is the basis for the "Gaia hypothesis"—a strongly debated topic within scientific and philosophical circles.[11]

Lovelock's Gaia hypothesis continues to be relevant in the scientific arena because of its relationship to the ongoing discussion and controversy about climate change*—the large-scale, long-term shift in the planet's weather patterns or average temperatures that, according to the consensus, human action has provoked. A search for the phrase "Gaia hypothesis" on the scientific database ScienceDirect* returns 373 scientific articles and 146 books. Additionally, a search for the phrase "Gaia hypothesis" on Google search engine returns over 400,000 pages on the Internet.

Why Does *Gaia: A New Look at Life on Earth* Matter?

Lovelock's Gaia hypothesis has been influencing academics, scientists, politicians, and the general public since the 1980s. While the idea of a self-regulating, living Earth was initially rejected by many scientists, so was the German scientist Alfred Wegener's* theory that the Earth's continents drift and move position over the planet's surface ("continental drift")* and the English naturalist

Charles Darwin's theory of evolution by means of natural selection, explaining how environment, adaptation, and inherited characteristics lead to the formation of new species. Both are accepted as fact today.[12]

Initially, Lovelock's hypothesis was severely criticized by geologists, evolutionary biologists* (those studying the Earth's living things in the light of evolutionary theory), and planetary scientists* (those engaged in the scientific study of planets and moons). For example, the evolutionary biologist Richard Dawkins* argued that organisms could not act as a whole to control their environment.[13]

Despite the widespread criticism, many scientists thought the idea was worth discussing further, and Lovelock's Gaia hypothesis started to receive more favorable attention in the late 1980s.[14] In 1985, the University of Massachusetts hosted the first public symposium on the Gaia hypothesis, titled "Is the Earth a Living Organism?"; in 1988, the American Geophysical Union* (AGU)'s First Chapman Conference on the Gaia Hypothesis was held in San Diego, California.[15] "Change is in the air," Lovelock wrote in 1994 when a scientific meeting on "The Self-regulating Earth" was held in the English city of Oxford. In 2000, the AGU held a second conference in Valencia, Spain;[16] in 2006, an international conference on the Gaia hypothesis was held at George Mason University in Virginia.[17]

In response to the initial criticism, Lovelock and his principal collaborators, the US geoscientist* Lynn Margulis* and the British marine and atmospheric scientist Andrew Watson,* developed the Gaia hypothesis into a theory that might be scientifically tested ("geoscience" is the branch of science drawing on geology, physics, and chemistry to consider the defining attributes of the Earth). Although the terms "Gaia hypothesis" and "Gaia theory" are often used interchangeably, Lovelock used them with separate meanings. For him, the hypothesis became a theory only after there began to be scientific evidence to support it—in his *Ages of Gaia* (1995), he writes that "we can now begin to think of Gaia as a theory, something rather more than mere 'let's suppose' of a hypothesis."[18]

"To understand even the atmosphere, the simplest of the planetary compartments," Lovelock has pointed out, "it is not enough to be a geophysicist;* knowledge of chemistry and biology is also needed"[19] (geophysics is the study of the various gravitational, magnetic, electrical, and seismic phenomena such as earthquakes that define our planet). The core ideas of the Gaia hypothesis were so broad and varied that it was not apparent immediately which major scientific discipline could accommodate them. So Lovelock proposed that such a broad topic had to be discussed under a new interdisciplinary branch of science (that is, a branch of science drawing the aims and methods of different fields of scientific inquiry).

The Gaia hypothesis ultimately gave birth to a new academic discipline known as geophysiology*—a field, also known as Earth system science, that studies the interaction of the Earth's living things and the Earth itself. Many top-ranked academic institutions around the world, including Stanford University in California, have established Earth system science departments to study the planet's oceans, lands, and atmosphere as an integrated system.

1. Ian Irvine, "James Lovelock: The Green Man," The *Independent*, December 3, 2005, accessed October 10, 2013, http://www.independent.co.uk/news/ people/profiles/james-lovelock-the-green-man-517953.html.
2. James E. Lovelock, *Homage to Gaia: The Life of an Independent Scientist*, rev. ed. (London: Souvenir Press Ltd., 2014), 1.
3. James E. Lovelock, *The Vanishing Face of Gaia: A Final Warning* (London: Penguin Books, 2010), 206.
4. Robin McKie, "Gaia's Warrior," *Green Lifestyle Magazine*, July/August 2007, 60–62.
5. "Curriculum Vitae," James Lovelock's official website, accessed December 29, 2013, http://www.jameslovelock.org/page2.html.
6. Peter Forbes, "Jim'll Fix it," The *Guardian*, February 21, 2009, accessed December 23, 2013, http://www.theguardian.com/culture/2009/feb/21/ james-lovelock-gaia-book-review.
7. Mary Midgley, "Great Thinkers—James Lovelock," *New Statesman*, 14 July 2003.
8. James E. Lovelock and C. E. Giffin, "Planetary Atmospheres: Compositional and Other Changes Associated with the Presence of Life," *Advances in the Astronautical Sciences*, 25 (1969): 179–93.
9. See James E. Lovelock, "Preface," in *Gaia: A New Look at Life on Earth*, by James E. Lovelock, rev. ed. (Oxford: Oxford University Press, 2000), ix.
10. W. Ford Doolittle, "Is Nature Really Motherly?," *CoEvolution Quarterly* 29 (1981): 58–65.
11. Richard R. Wallace and Bryan G. Norton, "Policy implications of Gaian Theory," *Ecological Economics* 6 (1992): 103.
12. Martin Ogle, "The Gaia Theory: Scientific Model and Metaphor for the 21st Century," *Revista Umbral* (*Threshold Magazine*) 1 (2009): 99–106.
13. See Richard Dawkins, *The Extended Phenotype: The Gene as the Unit of Selection* (Oxford: Oxford University Press, 1982), 234–36.
14. Lovelock, *Gaia: A New Look at Life on Earth*, xii.
15. Eric G. Kauffman, "The Gaia Controversy: AGU's Chapman Conference," *Eos, Transactions of the American Geophysical Union* 69 (1989), 763–64.
16. Brent F. Bauman, "The Feasibility of a Testable Gaia Hypothesis" (BSc thesis, James Madison University, 1998).
17. "Gaia Theory Conference at George Mason University," Arlington County, accessed December 27, 2013, http://www.gaiatheory.org/2006–conference/.
18. James E. Lovelock, *The Ages of Gaia: A Biography of Our Living Earth*, rev. ed. (Oxford: Oxford University Press, 1995), 44.
19. James E. Lovelock, "Geophysiology, the Science of Gaia," *Reviews of Geophysics* 27 (1989): 222.

SECTION 1
INFLUENCES

MODULE 1

THE AUTHOR AND THE HISTORICAL CONTEXT

KEY POINTS

- *Gaia: A New Look at Life on Earth* explains the concept of a self-regulating* Earth, a concept that has remained an original and important contribution to scientific debate around sustaining life on our planet; "self-regulating" describes the interactions of the Earth's organisms, air, water, and rocks in order to keep the planet fit and comfortable for life.
- Lovelock, an independent scientist through much of his career, says that the ability to think freely was critical to the development of the Gaia hypothesis,* according to which "the physical and chemical condition of the surface of the Earth, of the atmosphere, and of the oceans, has been and is actively made fit and comfortable by the presence of life itself."[1]
- Lovelock's short-term employment with NASA's* Jet Propulsion Laboratory in California on the unmanned Viking Mission to Mars* left him fascinated by the study of life on Earth and other planets (NASA is the organization responsible for the United States' civilian space program).

Why Read This Text?

James E. Lovelock's *Gaia: A New Look at Life on Earth* (1979) is one of the best-selling popular books on a contemporary scientific debate concerning life and planetary processes. In it, he describes a self-regulating, living Earth that he calls Gaia—after the ancient Greek* goddess of the Earth. Lovelock's radical Gaia hypothesis was fiercely criticized by evolutionary biologists,* chemists* (those with a specialist knowledge of the properties and reactions of matter), and planetary scientists* (those who study planets and moons). Even so, it has been proposed that the hypothesis has since been developed into a scientifically testable theory—the Gaia theory.*[2]

Over the last three decades, the Gaia hypothesis has been the subject of numerous scientific papers and books, and several international conferences.[3] Scientists and researchers from around the world recognized in the Amsterdam Declaration on Global Change* of 2001 that "the Earth system behaves as a single, self-regulating system comprised of physical, chemical, biological, and human components."[4] Although the Declaration does not mention Gaia, this is the key idea of the hypothesis.

The development by the Intergovernmental Panel on Climate Change (IPCC)* of Earth system models* to project future climate conditions has been a major achievement of the Gaia theory.[5] The IPCC is an intergovernmental body led by the United Nations;* Earth system models are used to evaluate regional and global climate under a number of different conditions by considering interactions between atmosphere, ocean, land, ice, and biosphere.*

> *"Humanity and science were offered a cornucopia of benefits from the accelerated inventions of the Second World War. Had we been less combative animals we could have used this new knowledge constructively. We could have made the observation of the Earth from space a priority, built satellites that viewed the land, the air, and the oceans, and seen the looming dangers of global warming in time; instead we made space missiles."*
>
> ——James E. Lovelock, *A Rough Ride to the Future*

Author's Life

James E. Lovelock was born on July 26, 1919 in Letchworth Garden City, England, into a working-class* family.[6] His father, Thomas A. Lovelock* had an interest in painting;[7] his mother, Nellie A. Elizabeth,* worked as a secretary. In his autobiography *Homage to Gaia*, Lovelock writes that "the bitterest blow for [my mother] came when she won a rare scholarship ... to a grammar school [a selective state school]. She could not take it because the family needed her earning power at thirteen to survive. Instead of an enlightened education ... she spent her days in a pickle factory sticking labels on the jars."[8] She was "full of working-class good intentions," he continues, "and she had an unquestioning belief in education. She was determined that I should go to a grammar school and as soon as possible. She had been denied the chance of a 'good education' and she did not intend that I should suffer from a lack of it."[9] While Lovelock did not particularly enjoy his school life, he was determined to become a scientist.

Lovelock started his professional career at the National Institute for Medical Research* in London. He worked there for nearly 20 years, with some career breaks. Never wanting to become a permanent employee of any educational or research organization, he became a fully independent scientist in the early 1960s; he has not been formally associated with any major university or research facility since then and has practiced science independently using the revenue earned from his inventions and publications. He writes that "one of the joys of independence is the extent to which the needs of different customers are shared in common: work done for one agency like NASA, often cross-fertilized the work I did for another, such as [the multinational oil and gas company] Shell."[10] Lovelock said this independence helped him tremendously in developing the radical Gaia hypothesis.

Author's Background

Lovelock's childhood, educational background, real-world experience, and many of his research collaborators immensely influenced his unusual way of thinking about life on Earth. From his childhood onward, he was particularly interested in nature and science. According to the *Green Lifestyle Magazine*,[11] his passion for science began with trips to science and natural history museums in London and with reading stories by the science fiction writers H. G. Wells,* the English author of *The War of the Worlds* (1898), and Jules Verne,* the French author of *From Earth to the Moon* (1865). Lovelock's own family greatly shaped his way of thinking and caring for nature. His greatest influence was his beloved father who expressed a great love for and care of nature throughout his life.[12] Lovelock describes in the book that his father used to say that every living creature on the surface of the Earth serves a certain purpose, and together they form a greater ecosystem* (a biological system made up of all the organisms found in a specific place) in which the physical environment interacts with living things.

Lovelock's quest for Gaia began in the early 1960s when he was working at the National Aeronautics and Space Administration (NASA) Jet Propulsion Laboratory on the Viking Mission* for exploring life on Mars. The unmanned space exploration was based on the theory that any evidence discovered for life on Mars would be similar to that for life on Earth. Lovelock proposed that a simple examination of the Martian atmosphere could tell if there was any life there. Soon, he became especially interested in and engaged with the subject of what life, precisely, is, and was motivated to conduct further research on Earth's atmosphere, the composition of oceans, and the role of living organisms* in the framework of planetary processes.

Lovelock's idea of a self-regulating Earth came to him in a time when there was no public concern about global climate change* (long-term change in the planet's weather patterns and local and global temperatures) or biodiversity* (the wide variety of life on the planet). His Gaia hypothesis offered an explanation for how greenhouse gases* (gases that trap the Sun's energy), protected early life-forms on Earth by keeping the atmosphere warm and comfortable; "the dangers of habitat destruction and inflation of the air with greenhouse gases," he has written, "seemed remote and trivial in the 1970s and 1980s."[13]

---

1. James E. Lovelock, Gaia: A New Look at Life on Earth, rev. ed. (Oxford: Oxford University Press, 2000), 144.
2. James W. Kirchner, "The Gaia Hypothesis: Conjectures and Refutations," Climatic Change 58 (2003): 21.
3. "Gaia Hypothesis," Environment website, accessed December 23, 2013, http://www.environment.gen.tr/gaia/70-gaia-hypothesis.html.
4. James E. Lovelock, "The Living Earth," Nature 426 (2003): 769–70.

5. James E. Lovelock, *A Rough Ride to the Future* (London: Penguin Books, 2015), 94.
6. Ian Irvine, "James Lovelock: The Green Man," The Independent, December 3, 2005, accessed October 10, 2013, http://www.independent.co.uk/news/ people/profiles/james-lovelock-the-green-man-517953.html.
7. James E. Lovelock, *Homage to Gaia: The Life of an Independent Scientist*, rev. ed. (London: Souvenir Press Ltd., 2014), 7–10.
8. Lovelock, *Homage to Gaia*, 7–37.
9. Lovelock, *Homage to Gaia*, 15–16.
10. Lovelock, *Homage to Gaia*, 282.
11. Robin McKie, "Gaia's Warrior," Green Lifestyle Magazine, July/August 2007, 60–62.
12. Lovelock, *Homage to Gaia*, 8.
13. See James E. Lovelock, "Preface," in Gaia: A New Look at Life on Earth, viii.

MODULE 2

ACADEMIC CONTEXT

KEY POINTS

- When Lovelock worked at the United States space agency NASA,* he was ridiculed for his ideas about testing for life on Mars, and the scientific community initially ignored his Gaia hypothesis* (according to which the surface of the Earth and the life it supports are a self-regulating* entity).
- Lovelock was unaware that earlier scientists had briefly discussed the idea of a self-regulating Earth when he began to develop the Gaia hypothesis ("self-regulating" here means that the mutual interactions of the Earth's organic and nonorganic features keep the planet fit and comfortable for life).
- With one of his students, Lovelock created a mathematical model of an environment he called Daisyworld* to test his ideas.

The Work in Its Context

James E. Lovelock's *Gaia: A New Look at Life on Earth* is an original contribution to the understanding of the atmosphere and the role of organisms* in influencing the Earth's climate. The late 1950s and the 1960s were a new era in space exploration.* In 1957, the Soviet Union* launched Sputnik,* the first ever artificial satellite* to orbit the Earth.[1] The United States had a long rivalry with the Soviet Union and, in 1961, President John F. Kennedy* began expanding the US space program through NASA. He set a target to land a man on the Moon and return him safely home by the end of the decade.[2]

NASA was also preparing to launch spacecraft to Mars to

search for life. In those days, little research was conducted on life beyond Earth; biologists* (those engaged in the scientific study of organisms) designed experiments to try to copy the conditions on Mars, but these experiments had to be conducted on Earth. Lovelock, employed at NASA, was skeptical about experiments conducted on Earth to detect life on Mars; as an alternative to sampling Mars's soil to detect any presence of life, he proposed sampling the planet's atmosphere. His idea was rejected and ridiculed at NASA.[3]

Similarly, he failed to stimulate interest in the scientific community when he first presented his Gaia hypothesis at a conference in the 1960s. The idea of a self-regulating Earth was complex and could not be placed in any single academic field. It was not until he published *Gaia* in 1979 that the academic community took interest in his hypothesis.

> *"I expected to discover somewhere in the scientific literature a comprehensive definition of life as a physical process, on which one could base the design of life-detection experiments, but I was surprised to find how little had been written about the nature of life itself."*
>
> —— James E. Lovelock, *Gaia: A New Look at Life on Earth*

Overview of the Field

Although Lovelock's self-regulating Earth was not discussed in any formal academic discipline before he introduced the idea,[4] as he writes,"the idea that the Earth is alive in a limited sense is probably as old as humankind."[5] Several great scientists in the past looked at the planet Earth as a living body. In 1785, James Hutton,* a famous Scottish geologist,* described Earth as a self-regulating system. Hutton, who said Earth was like a living creature, compared the cycling of Earth's nutrients (food) between soil and plants, and the movement of water from oceans to land and back, with the circulation of blood in a human body.[6] In *The Ages of Gaia* (1995), Lovelock writes, "James Hutton is rightly remembered as a deeply influential figure in the field of geology but his idea of a living Earth was forgotten, or denied, in the intense reductionism* of the nineteenth century"[7] ("reductionism" occurs when we see something complex as merely the sum of its parts, without considering how those parts may interact with each other). In the early twentieth century the Russian scientist Vladimir I. Vernadsky* discussed the notion of a living Earth with an envelope of life and popularized the term "biosphere"* to describe the area of living matter.[8]

Lovelock, however, was unaware of these ideas when he formulated his hypothesis, which developed the idea of a self-regulating Earth more than anyone had previously done. Lovelock's Gaia hypothesis was considered by many scientists as contradictory to the deeply influential English naturalist Charles Darwin's* theory of evolution,* which suggested that organisms evolve but their nonliving environment does not. Later, however, Lovelock demonstrated that living organisms could interact with their nonliving environments to achieve the best living conditions for themselves.[9]

Academic Influences

Lovelock's childhood experiences and relationship with his parents, his education, his experience in the working world, and his collaborators among research scientists all had a major influence on the unusual perspective he developed in thinking about life on Earth. He was also influenced by several early scientists; in his autobiography, he wrote that he was greatly influenced by the English American zoologist* George E. Hutchinson's* work on the Earth's biochemistry* (a zoologist is engaged in the scientific study of animals; biochemistry is the study of the chemical processes of living things). Hutchinson, who studied the interaction of organisms and their environment, described the Earth as a self-regulating body from the viewpoint of chemical activity.[10] Lovelock also wrote that his friend Sidney Epton,* a chemist, helped stir the first real public interest in Gaia by coauthoring an article in the popular science journal *New Scientist* in 1975.[11]

The concept of a self-regulating Earth and the development of the Gaia hypothesis were greatly shaped and augmented by Lovelock's main collaborators—the geoscientist* Lynn Margulis,* the philosopher Dian Hitchcock* and Epton. Lovelock and Margulis started working together on the Gaia hypothesis in the early 1970s. Lovelock describes this partnership as a "most rewarding scientific collaboration" in his book that led to the publication of their first joint scientific paper on the topic.[12] Margulis was not only a research collaborator but a good friend who believed in Lovelock so much that she helped him secure funding for research on the Gaia hypothesis and for publication of his second book, *The Ages of Gaia* (1995).[13] Later on, Andrew Watson,* a British PhD candidate, worked with Lovelock on his research. Watson and Lovelock together developed a mathematical model of an imaginary Earthlike planet, "Daisyworld,"*[14] in an effort to demonstrate that living organisms can actually control the atmosphere and climate in which they live.

---

1. James E. Lovelock, *The Ages of Gaia: A Biography of Our Living Earth*, rev. ed. (Oxford: Oxford University Press, 1995), 4.
2. "Space Program," John F. Kennedy Presidential Library and Museum, accessed January 8, 2016, http://www.jfklibrary.org/JFK/JFK-in-History/Space-Program.aspx.
3. James E. Lovelock, *Homage to Gaia: The Life of an Independent Scientist*, rev. ed. (London: Souvenir Press Ltd., 2014), 250.
4. Lovelock, *The Ages of Gaia*, 3–14.
5. Lovelock, *The Ages of Gaia*, 9.
6. Lovelock, *The Ages of Gaia*, 9.
7. Lovelock, *The Ages of Gaia*, 9.
8. Lovelock, *The Ages of Gaia*, 10.
9. Lovelock, *The Ages of Gaia*, 41–61.
10. Lovelock, *Homage to Gaia*, 263.
11. James E. Lovelock and Sidney Epton, "The Quest for Gaia," *New Scientist* 65, no. 935 (1975): 304–09.
12. James E. Lovelock and Lynn Margulis, "Atmospheric Homeostasis by and for the Biosphere: The Gaia Hypothesis," *Tellus* 26, nos. 1–2 (1974): 1–10.
13. Lovelock, *Homage to Gaia*, 369.
14. Andrew J. Watson and James E. Lovelock, "Biological Homeostasis of the Global Environment: The Parable of Daisyworld," *Tellus* 35B (1983): 286–89.

MODULE 3
THE PROBLEM

KEY POINTS

- In the mid-1960s when Lovelock conceived the idea of a self-regulating Earth,* researchers at the United States space agency NASA* and academics were primarily interested in detecting life on other planets by using well-established biological experiments that were only tested on Earth.
- Lovelock's Gaia hypothesis* holds that every organism* on Earth is tightly coupled with its environment, and that the sum of these interactions maintains a suitable living condition.
- Earlier scientists had suggested in passing that Earth is a self-regulating, living system, but Lovelock devoted much of his life to trying to prove it.

Core Question

James Lovelock's *Gaia: A New Look at Life on Earth* should be read in the light of the author's preoccupation with the questions "What is life?" and "How should it be recognized?"

These core questions, although simple in nature, were indeed original and critical for the development of Lovelock's ideas of the evolution of life and the Earth as a single living entity. He spent many years developing his radical and groundbreaking idea into the Gaia hypothesis that every living organism on Earth—ranging from a tiny, microscopic virus to the largest whale—can be regarded as tightly coupled with its environment as a single entity. These coupled entities are capable of maintaining a suitable living condition by controlling the Earth's atmosphere and the climate.

Lovelock's formulation of the hypothesis was influenced by his time at NASA's Jet Propulsion Laboratory in the mid-1960s. There, he found himself fascinated by the images of Earth taken by the astronauts on the agency's manned Apollo* space missions, which, he writes, led him to look at Earth's surface from the top down rather than from the bottom up. The thought of a living planet came to him while working on experiments for detecting life on other planets. He thought there must be a planet-scale regulation system that has always kept the Earth fit and comfortable for life, while Earth's closest neighbors, Mars and Venus, were lifeless.[1] He challenged his fellow scientists at NASA on the usefulness of conducting experiments for detecting life on Mars from the soil, proposing as an alternative that experiments to determine the composition of Mars's atmosphere would indicate the presence or absence of life.[2]

> *"... [T]hinking about life on Mars gave some of us a fresh standpoint from which to consider life on Earth and led us to formulate a new, or perhaps revive a very ancient, concept of the relationship between the Earth and its biosphere."*
>
> ——James E. Lovelock, *Gaia: A New Look at Life on Earth*

The Participants

In the mid-1960s, when Lovelock conceived the idea of a living Earth, there was no ongoing debate about this topic. Space exploration was at its peak, fueled by the rivalry between the Soviet Union* and the United States—ideologically opposed superpowers, then vying for global military and cultural dominance. At NASA, biologists* and soil scientists were designing life-detecting experiments for spacecraft; these experiments, however, were conducted on the surface of Earth. One of the scientists was Vance Oyama,* a biochemist* who insisted that Martian soil should be collected and examined for the presence of life. At that time, the definition of "life" and whether it can control the living environment on Earth's surface were not discussed in any traditional academic discipline.[3]

Lovelock's concept of a living Earth, Gaia, is closely linked to the concept of life; indeed, to understand Gaia, one must understand life. When he started his in-depth research on life and its significance on Earth's environment, Lovelock found that several early scientists had suggested that the Earth was alive. In 1785, the Scottish geologist* James Hutton,* known as the father of geosciences,* proposed that Earth was a superorganism*—a living entity composed and defined through the mutual interactions of the organisms inhabiting it. In the late nineteenth century, the English biologist Thomas H. Huxley* proposed that the planet Earth was a living, self-regulating system, while the Russian geochemist* Vladimir I. Vernadsky* suggested that life had shaped the Earth as an active geological force.[4] None of these early scientists, however, looked any further than their initial thoughts. Lovelock, however, took a fresh approach to this idea and made it his life's work. He created the Gaia hypothesis that explains how the Earth maintains a condition that is always fit and comfortable for life. This stability in the midst of constant change is called "homeostasis."*

The Contemporary Debate

The idea of a living Earth, although conceived by previous scientists, was not accepted by mainstream scientists, and was unknown outside a few old scientific publications. The Ukrainian philosopher and independent scientist Yevgraf M. Korolenko,* for instance, declared the Earth to be a living organism near the end of the nineteenth century.[5] However, when Lovelock himself conceived the idea of a living Earth and started asking the same questions he was not aware of these old statements or publications. There was no ongoing debate and no publications directly useful to answering the central questions of the self-regulating Earth.

The subject matter was so broad that Lovelock needed advice and collaboration to develop his hypothesis into a theory that was capable of making testable scientific predictions.[6] While many scientists simply criticized his idea, some agreed to collaborate with him: Dian Hitchcock,* Sidney Epton,* and Lynn Margulis,* particularly, actively helped him develop the Gaia hypothesis.

---

1. James E. Lovelock, *Gaia: A New Look at Life on Earth*, rev. ed. (Oxford: Oxford University Press, 2000), 1–29.
2. James E. Lovelock, *Homage to Gaia: The Life of an Independent Scientist*, rev. ed. (London: Souvenir Press Ltd., 2014), 242–43.

3. James E. Lovelock, *The Ages of Gaia: A Biography of Our Living Earth*, rev. ed. (Oxford: Oxford University Press, 1995), 15–20.
4. See Crispin Tickell, "Foreword," in *The Revenge of Gaia: Why the Earth is Fighting Back—and How We Can Still Save Humanity*, by James E. Lovelock (London: Penguin Books, 2007), xiv.
5. Lovelock, *The Ages of Gaia*, 8–10.
6. Lovelock, *The Ages of Gaia*, 41–61.

MODULE 4

THE AUTHOR'S CONTRIBUTION

KEY POINTS

- Although Lovelock's main aim in writing the book was to dconvey the core idea of a self-regulating,* living Earth to a general audience, he was aware that scientists might read about his Gaia hypothesis.*
- Lovelock's controversial Gaia hypothesis has changed the way life is viewed on Earth and eventually gave rise to the new academic field of geophysiology* or Earth system science*—a discipline founded on the principle that the Earth can be understood as a system of interactions in which life plays a significant role.
- It took more than a decade to develop the Gaia hypothesis beyond the mere idea of a self-regulating Earth; after discussions with scientists from several disciplines and with the help of a few close collaborators, the idea became the Gaia theory.*

Author's Aims

James E. Lovelock started writing *Gaia: A New Look at Life on Earth* in 1974. He believed that all living beings, including humans, are part of a community that is unconsciously keeping the Earth a comfortable place. He even started to feel that humans are like any other living organisms, with no particular rights but only obligations to the community of Gaia.[1] Lovelock wanted to share these ideas widely with the general public, not just with scientists who were not open to these rather controversial ideas at that time.

Lovelock conceived the idea of Gaia in 1965 while working at NASA.* He first published the idea in 1967 in the international journal *Icarus*,* and then put the idea before his fellow scientists in 1971 in a presentation with the title "Gaia as Seen Through the Atmosphere."[2] Prestigious mainstream journals such as *Science* and *Nature*, however, were not ready to accept any paper on the Gaia hypothesis; "the very idea of detecting life on a planet by atmospheric analysis," Lovelock writes, "must have seemed outrageous to the conventional astronomers* and biologists* who reviewed our paper."[3]

In 1979, Lovelock finally published *Gaia*—a collection of all the ideas up to that point. He knew that his idea was significant and that he needed a concrete plan to translate the hypothesis into a testable scientific theory. Although he discussed the ideas with many scientists over the next decade, only a few supported his ideas and collaborated with him. Lovelock wanted to show his fellow scientists that Earth needed to be seen from top down, not the other way around; one can only see from space (he argued) that, compared to its dead neighbors, our blue planet Earth is living.

> *"The idea of the Earth as a kind of living organism, something able to regulate its climate and composition so as always to be comfortable for the organisms that inhabited it, arose in a most respectable scientific environment. It came to me suddenly one afternoon in 1965 when I was working at the Jet Propulsion Laboratory (JPL) in California."*
>
> ——James E. Lovelock, *Gaia: A New Look at Life on Earth*

Approach

Lovelock writes his book, *Gaia: A New Look at Life on Earth* as a story of the discovery of life on the planet Earth. He begins the book by describing NASA's* space mission to search for life on other planets of the solar system. Lovelock addressed in an original way the core question of how to detect life on a distant world like Mars or Venus, suggesting that instead of testing soil, a planet's atmosphere can provide the necessary evidence if life indeed exists on its surface. This controversial idea departed from the scientific orthodoxy of that time, and Lovelock knew that without any evidence no one would believe him.

Between 1965 and 1975, he gathered as much supporting information as possible to develop his individual ideas into a working hypothesis. He thought the best way to describe his hypothesis was to write a book as a story of discovery. So in a series of interlinked chapters, Lovelock describes the atmosphere of the early Earth and how life evolved from tiny, single-celled organisms to more complex forms such as humans. Lovelock explains how living organisms interact with their nonliving environment to form a self-regulating entity, which he called Gaia. No author had previously offered such an in-depth analysis of the intricate yet complementary relationship between living and nonliving components within Gaia.

Contribution in Context

Although the core concept of a self-regulating, living Earth was not new, Lovelock's approach to researching it was original. At that time, no one was engaged in any research on the evolution of life and the evolution of the Earth as a single entity. When Lovelock started detailed research into the existing literature on life and its relationship with its environment, he could not find any. However, he stumbled upon some old notes and literature in which authors suggested that the Earth was alive. In the eighteenth century, for example, the geologist* James Hutton* said that the Earth was like an animal. A century later, the Austrian geologist Eduard Suess* introduced the word "biosphere",* which was further developed by the Russian geochemist* Vladimir I. Vernadsky,* who suggested that it can be regarded as the area on Earth in which energy is produced that sustains life.[4] While Lovelock was entirely ignorant of the related ideas of these earlier scientists, he later acknowledged those pioneers, referencing them in his second book *The Ages of Gaia* (1995).

---

1. James E. Lovelock, *Gaia: A New Look at Life on Earth*, rev. ed. (Oxford: Oxford University Press, 2000), ix.
2. James E. Lovelock, *The Ages of Gaia: A Biography of Our Living Earth*, rev. ed. (Oxford: Oxford University Press, 1995), 8.
3. James E. Lovelock, *Homage to Gaia: The Life of an Independent Scientist*, rev. ed. (London: Souvenir Press Ltd., 2014), 250.
4. Lovelock, *The Ages of Gaia*, 8–10.

SECTION 2
IDEAS

MODULE 5

MAIN IDEAS

KEY POINTS

- A key part of the Gaia hypothesis* is a well-balanced and stable living condition known as homeostasis,* which results from interaction between living things and the environment.
- Homeostasis is maintained by a "cybernetic feedback mechanism"*—an automatic control system that makes adjustments in response to change.
- Lovelock warned that humans could break the feedback mechanism that sustains Gaia through too much change to the atmosphere and surface of the Earth.

Key Themes

James E. Lovelock's *Gaia: A New Look at Life on Earth* (1979) is a story about a search for life. Central to this book's story, the Gaia hypothesis suggests that the Earth constantly maintains the interaction of its physical, chemical, and biological environments so that organisms live and survive, evolving over time. This relative stability, called homeostasis, is fundamental for understanding the Gaia hypothesis.

In *Gaia*, Lovelock writes that the Earth maintains homeostasis through an automatic system called a cybernetic feedback mechanism.*[1] This system controls the temperature and the makeup of the Earth's atmosphere and oceans by making adjustments in response to change. The book says when the Earth's infant atmosphere was formed around 3.5 billion years ago, Gaia—the living Earth—shielded life on the planet by keeping the atmosphere suitable for living organisms.* The atmosphere also protected the planet from damaging cosmic radiation*—waves of energy from space—and from bombardment by meteorites (rocks from space).[2]

The final theme of the book is sustainable living within Gaia. Lovelock describes the relationship between humans and Gaia and warns us of dire consequences if the Earth's internal feedback mechanism fails. Such a failure could occur if humans cause excessive modifications of the Earth's surface and atmosphere. The book describes how humans are burning oil, gas, and coal more than ever before, which adds more of the gas carbon dioxide* to the atmosphere. This contributes to global warming* (an increase in global temperatures) because of the greenhouse effect: * the higher concentration of carbon dioxide and other greenhouse gases* in the atmosphere traps more heat from the Sun.[3]

> *"Earth's surface temperature is actively maintained comfortably for the complex entity which is Gaia, and has been so maintained for the most of her existence."*
>
> ——James E. Lovelock, *Gaia: A New Look at Life on Earth*

Exploring the Ideas

In *Gaia*, Lovelock explains how the composition of Earth's atmosphere has changed over millions of years. Various atmospheric gases, primarily nitrogen and oxygen, combined to create a homeostatic condition (roughly, an equilibrium),* in which life is sustained despite a constantly changing environment. Lovelock argues that Earth is different from its neighboring planets.[4] The atmospheres of Mars and Venus are primarily carbon dioxide gas and the life-supporting oxygen gas is almost absent. In contrast, Earth's atmosphere has about 21 percent oxygen, which is crucial for living animals and plants. Lovelock uses a conceptual model to demonstrate that on an Earthlike planet that has no life and has reached a state of chemical equilibrium, only a trace amount of life-sustaining oxygen would be present.[5] But the oxygen content in the Earth's atmosphere has remained about 21 percent for the past 200 million years.[6] The consistent oxygen concentration suggests that an active control system is in place on the Earth's surface.

Lovelock calls this control system a cybernetic feedback mechanism.* To explain the concept, he gives examples of household appliances such as an electric oven, an iron, and a room heater. These appliances are equipped with a thermostat* that controls the desired temperature by switching on and off. A certain temperature is maintained through a feedback mechanism that tells the device to switch off when the temperature rises above the desired level. Similarly, the human body has feedback mechanisms to maintain its internal temperature, so that our bodily functions can keep us alive.[7] Lovelock emphasizes that a cybernetic feedback mechanism may exist at a planetary scale on Earth, where plants and animals have the capacity to regulate the Earth's climate.[8]

In the last two chapters, Lovelock turns to the theme of living sustainably within Gaia. He warns the reader that if humans keep modifying the surface and the atmosphere of the Earth then the delicate homeostatic system of the planet—which took several billion years to develop—will ultimately collapse and humans will face severe consequences. Lovelock expresses his great fear that the coexistence of life and Gaia is currently at stake because humans, for their own comfort, are continuously modifying the Earth's landscape and its delicate atmosphere. He writes in the book that "changes in the production rates of greenhouse gases may cause perturbations on the global scale"[9] that may seriously interfere with Gaia's state of homeostasis, and ultimately could threaten the existence of life on Earth.[10]

Language and Expression

In *Gaia*, Lovelock writes for a general audience, using simple language so that nonscientists can easily understand the strange and unfamiliar idea of a self-regulating, living Earth:"I tried to write this book so that a dictionary is the only aid needed."[11] According to the popular science journal *New Scientist*,"Lovelock writes beautifully. A book that is both original and well written is indeed a bonus."[12]

Lovelock has been successful; his simple but effective method ultimately brought him fame as a popular science writer. But on account of his simple writing style, lack of evidence, and the controversial nature of the topic, many scientists strongly criticized the book. The nonscientific nature of the book was emphasized by Lovelock's decision to name the self-regulating planet Gaia after the ancient Greek goddess of Earth[13] (a name suggested by the British novelist William Golding, author of *Lord of the Flies*, who was a resident of Lovelock's village).[14]

Despite the problems he had getting the idea recognized by scientists, Lovelock later said he never regretted the choice of name.

---

1. See James E. Lovelock, "Cybernetics," in *Gaia: A New Look at Life on Earth*, by James E. Lovelock, rev. ed. (Oxford: Oxford University Press, 2000), 44–58.
2. Lovelock, *Gaia: A New Look at Life on Earth,* 59–77.
3. Lovelock, *Gaia: A New Look at Life on Earth,* 100–32.
4. James E. Lovelock, *Homage to Gaia: The Life of an Independent Scientist*, rev. ed. (London: Souvenir Press Ltd., 2014), 244.
5. Lovelock, *Gaia: A New Look at Life on Earth*, 30–46.
6. James E. Lovelock, *The Ages of Gaia*, rev. ed. (Oxford: Oxford University Press, 1995), 124.
7. Lovelock, *Gaia: A New Look at Life on Earth*, 49.
8. Lovelock, *Gaia: A New Look at Life on Earth*, 58.
9. Lovelock, *Gaia: A New Look at Life on Earth*, 113.
10. James E. Lovelock, *A Rough Ride to the Future* (London: Penguin Books, 2015), 75–111.
11. Brent F. Bauman, "The Feasibility of a Testable Gaia Hypothesis" (BSc thesis, James Madison University, 1998), 8.
12. Kenneth Mellanby, "Living with the Earth Mother," *New Scientist* 84 (1979): 41.
13. Toby Tyrrell, *On Gaia: A Critical Investigation of the Relationship between Life and Earth* (Princeton: Princeton University Press, 2013), 2.
14. Lovelock, *Homage to Gaia*, 255.

MODULE 6
SECONDARY IDEAS

KEY POINTS

- Trapping the Sun's heat, greenhouse gases* kept the early Earth's surface warm and comfortable, sustaining delicate life when the Sun's heat was much less intense than now.
- Self-regulating* mechanisms keep the level of oxygen in the atmosphere and the level of salt in the ocean relatively constant.
- The name Gaia and talk of a living Earth may initially have diverted the scientific community from giving the Gaia hypothesis* the attention it later received.

Other Ideas

In James E. Lovelock's *Gaia: A New Look at Life on Earth* there are several secondary ideas important to the development of his groundbreaking Gaia hypothesis:

- How are concentrations of different gases in Earth's atmosphere controlled?
- What is the importance of chemical reactions*—the effect of the combination of different chemicals—to the sustaining of life on Earth?
- What was the role of greenhouse gases in protecting the early Earth?
- Why is the sea not more salty, and what controls the salinity of seawater?

These secondary ideas, here stated as questions, help Lovelock explain and expand on the main themes. He explains, for example, how a constant supply of energy from chemical reactions is needed if life is to be sustained; these reactions occur when two or more chemicals combine to create a different chemical. If a planet is in a state of chemical equilibrium* (balance), however, no reactions take place and no energy is produced.[1] Lovelock explains that "in such a world there is no source of energy whatever: no rain, no waves or tides, and no possibility of a chemical reaction which would yield energy."[2] Without the presence of any chemical reaction, no energy is produced on Mars or Venus—and these planets, our closest neighbors, are indeed lifeless.[3]

> *"The human species is of course a key development for Gaia, but we have appeared so late in her life that it hardly seemed appropriate to start our quest by discussing our own relationships within her."*
>
> ——James E. Lovelock, *Gaia: A New Look at Life on Earth*

Exploring the Ideas

Life on Earth probably began as a simple, single-cell form under unsettled conditions of cosmic bombardment and active radioactivity*[4] (radioactivity occurs when atoms decay and release radiation*—roughly, waves of energy). Because of a lack of oxygen in the atmosphere, the surface of the Earth was exposed to the Sun's ultraviolet* radiation, which is not visible to the human eye. The atmosphere of the early Earth was filled with the gases carbon dioxide* and ammonia* (a combination of nitrogen and hydrogen). Lovelock suggests that these greenhouse gases kept the planet warm 3.5 billion years ago, when the Sun gave off 25 percent less heat than it does today.[5] Lovelock argues that 25 percent less heat from the Sun would imply an average surface temperature of well below zero degrees, which would cover the Earth's surface with snow and ice. From the geological* records, however, it is known that for this period the climate of the Earth was never wholly unfavorable for life.[6] So while the Sun has steadily grown stronger over 3.5 billion years, Gaia has been sustaining life on Earth under a relatively stable climate.

Earth's atmosphere is made up primarily of nitrogen and oxygen, with small amounts of carbon dioxide and other minor gases. These gases, which play an important role in regulating the Earth's temperature and climate, are controlled by biota*—the animal and plant life in the environment. For instance, a constant level of oxygen in the atmosphere is possible because of an active biotic control. Green plants and algae* (a plantlike organism that lives in water) use light from the Sun to produce oxygen through a process known as photosynthesis,* creating nutrients and giving off oxygen in the process. This adds life-sustaining oxygen to the atmosphere—and yet, the level of oxygen remains constant at about 21 percent.

So how does Gaia control the Earth's atmosphere to keep an oxygen level suitable for life? The key to the control of oxygen is another gas, methane,* a combination of carbon and hydrogen, produced primarily by single-celled organisms—bacteria.* Although the atmosphere contains only trace amounts of methane, it is critical in controlling oxygen levels. Lovelock compares the function of methane with the function of glucose (a sugar) in a person's blood.[7] Glucose provides energy to cells in the human body, so maintaining fairly constant glucose levels is essential for healthy cells and, by extension, a healthy body. Similarly, methane controls the level of oxygen in the atmosphere by reacting with oxygen to form carbon dioxide and water.

Lovelock explains how the world's oceans are kept in balance. Rains and rivers dissolve salt from rocks and carry it over the land into the ocean, yet the salinity level in oceans has remained constant at around 3.4 percent for a very long time.[8] Lovelock argues that, throughout the history of the oceans' life, the salinity level could not have exceeded 6 percent as indicated by fossil records. At higher salinity levels, the life-forms in the ocean that we see today would have evolved very differently.[9] Lovelock argues that because salinity levels have been steady in the oceans, there must be a mechanism for the ocean to get rid of some of its salt.[10] He suggested that "excess salt accumulates in the form of evaporites* [deposits] in shallow bays, land-locked lagoons, and isolated arms of the sea, where the rate of evaporation* is rapid and the inflow [of salt] from the sea is one-way."[11]

Overlooked

Many scientists misinterpreted Lovelock's Gaia hypothesis. Lovelock writes in the revised edition that the book "is not for hard scientists. If they read it in spite of my warning, they will find it either too radical or not scientifically correct."[12] But scientists did read it as a scientific text, and reacted as Lovelock describes. The idea that living animals can change their environments through interactions with their nonliving counterparts was simply rejected by evolutionary biologists.* Other scientists called the Gaia hypothesis teleological* (in line with a belief that nature has a purpose) and goal-seeking*(based on repeated analyses looking for a particular answer). They argued that to self-regulate, organisms would need foresight and planning, which was impossible.[13]

Lovelock commented that "they see Gaia as metascience [beyond science], something like religious faith, and therefore from their deeply held materialistic beliefs, something to be rejected."[14] Two of Lovelock's supporters in the United States, the biologist* Stephen Schneider* and the environmental scientist Penelope J. Boston,* said in their book *Scientists on Gaia* (1993)[15] that the Gaia hypothesis wrongly attracted the most attention from theologians* who study religious ideas, usually through scripture. The turn of phrase "living Earth" and the name "Gaia" may have diverted the attention of the scientific community away from a serious analysis of the hypothesis and its implications.

---

1. See James E. Lovelock, "The Recognition of Gaia," in *Gaia: A New Look at Life on Earth*, by James E. Lovelock, rev. ed. (Oxford: Oxford University Press, 2000), 32.
2. Lovelock, *Gaia: A New Look at Life on Earth*, 33.
3. Lovelock, *Gaia: A New Look at Life on Earth*, 30–43.
4. See James E. Lovelock, "In the Beginning," in *Gaia: A New Look at Life on Earth*, by James E. Lovelock, rev. ed. (Oxford: Oxford University Press, 2000), 15.
5. Lovelock, *Gaia: A New Look at Life on Earth*, 18.
6. Lovelock, *Gaia: A New Look at Life on Earth*, 18.

7. Lovelock, *Gaia: A New Look at Life on Earth*, 67.
8. Kate Ravilious, "Perfect Harmony," The *Guardian*, April 28, 2008, accessed December 30, 2013, http://www.theguardian.com/science/2008/apr/28/ scienceofclimatechange.biodiversity.
9. Lovelock, *Gaia: A New Look at Life on Earth*, 86.
10. James E. Lovelock, *The Ages of Gaia: A Biography of our Living Earth*, rev. ed. (Oxford: Oxford University Press, 1995), 99–107.
11. Lovelock, *Gaia: A New Look at Life on Earth*, 91.
12. Lovelock, *Gaia: A New Look at Life on Earth*, xii.
13. James E. Lovelock, *Homage to Gaia: The Life of an Independent Scientist*, rev. ed. (London: Souvenir Press Ltd., 2014), 264.
14. Lovelock, *Gaia: A New Look at Life on Earth*, xii.
15. Stephen H. Schneider and Penelope J. Boston, eds., *Scientists on Gaia* (Cambridge: The MIT Press, 1993), 433.

MODULE 7
ACHIEVEMENT

KEY POINTS

- Lovelock wrote *Gaia: A New Look at Life on Earth* for a general audience because he believed, correctly, that the scientific community would not take it seriously.
- His warning about the danger of damaging the self-regulating* mechanisms of the Earth is directly related to today's concern about climate change.*
- The Gaia hypothesis* has led to the establishment of the interdisciplinary field of study called Earth system science.*

Assessing the Argument

James Lovelock's first book, *Gaia: A New Look at Life on Earth*, received a mixed reception when it was published in 1979. Lovelock later wrote in his autobiography that "its publication completely changed my life and the fall of mail through my letterbox increased from a gentle patter to a downpour, and has remained high ever since."[1] The main interest in Gaia came from the general public, philosophers, and religious leaders; only a third of the letters were from scientists.[2]

Although he did not write the book as a science text for specialists, Lovelock expected that some scientists would read it. Indeed they did—and many rejected the Gaia hypothesis. After the work's publication, Lovelock's scientific colleagues asked him why he reported the ideas about Gaia hypothesis in a book and not in peer-reviewed scientific journals.[3] He responded that when he published the first scientific article on Gaia in the early 1970s, the opposition, mainly from evolutionary biologists,* was so strong that he believed the editors of prestigious journals such as *Science* or *Nature* would reject such articles.[4]

The situation changed in the 1990s, however, and Gaia papers became easier to publish, even in prestigious journals. Lovelock commented that the early rejection by biologists and the conservatism of editors of some scientific journals helped him realize the potential of the hypothesis in stimulating later scientific debates and discovery.[5]

> *"Now most scientists appear to accept Gaia theory and apply it to their research, but they still reject the name Gaia and prefer to talk of Earth System Science, or Geophysiology, instead."*
>
> ——James E. Lovelock, *Gaia: A New Look at Life on Earth*

Achievement in Context

Lovelock's Gaia hypothesis was born in the mid-1960s, when the United States and the Soviet Union* were engaged in a serious battle over space exploration, meaning that there was a significant focus on space. His work at NASA* on scientific experiments for life detection on Mars led him to think about why Earth was different from its neighbors, Mars and Venus.

Since the work's publication in 1979, the Gaia hypothesis has been actively debated in scientific circles. One result of that debate was the development of a mathematical model called "Daisyworld,"*[6] which ecologists use to test the role of biodiversity* (the richness of species in a specific place) and stability of ecosystems* (habitat, the organisms that live in it, and the interactions between the two) in sustaining a healthy living environment.[7]

In his book, Lovelock also explains how the delicate self-regulating system of Earth can collapse as a result of excessive modification of the Earth's surface and atmosphere. In the 1970s, when the Gaia hypothesis predicted global changes, there was no evidence. Today, it is widely accepted that changes in atmospheric greenhouse gases* such as carbon dioxide are affecting the climate on a global scale.[8] These changes are called "anthropogenic,"* meaning they were caused by human activity. Lovelock warned about the danger of anthropogenic global warming* through the greenhouse effect* more than 30 years ago. The Gaia hypothesis is directly related to today's concerns about climate change.

Limitations

Gaia was published in 1979 in nonscientific and accessible language, and was received very well by the general public but rejected by some scientists, notably evolutionary biologists.[9] In addition to disagreeing on matters of science, scientists felt the storytelling approach and association of the book with a mythological character meant it could not be taken seriously.[10] Lovelock acknowledged that the "Gaia book was hypothetical, and lightly written—a rough pencil sketch that tried to catch a view of the Earth seen from a different perspective."[11] He also responded to criticisms from scientists by stating "I wrote this book when we were only just beginning to glimpse the true nature of our planet and I wrote it as a story of discovery."[12]

Another factor that limited the consideration of *Gaia* was that the very idea of a self-regulating Earth was so broad that it did not fit into a single, traditional academic discipline, such as geology,* biology,* or physics.*[13] At that time, taking an interdisciplinary approach to science was not yet common. Today, the Gaia hypothesis has moved from being "a metaphor,* not a mechanism,"[14] as the US evolutionary biologist Stephen J. Gould* understood it, to become the heart of a new interdisciplinary field of study: Earth system science. Lovelock's Gaia hypothesis was not limited to any particular time or place, and has inspired many academic disciplines, including ecology* (the branch of biology that looks at how organisms relate to one another and to their physical surroundings), marine biology* (the study of life in the oceans), and climate science.*[15]

---

1. James E. Lovelock, *Homage to Gaia: The Life of an Independent Scientist*, rev. ed. (London: Souvenir Press Ltd., 2014), 264.
2. Lovelock, *Homage to Gaia*, 264.
3. James E. Lovelock, *The Ages of Gaia: A Biography of Our Living Earth*, rev. ed. (Oxford: Oxford University Press, 1995), xiii–xxii.
4. Lovelock, *The Ages of Gaia*, xv.
5. Lovelock, *The Ages of Gaia*, xiv–xv.
6. Andrew J. Watson and James E. Lovelock, "Biological Homeostasis of the Global Environment: The Parable of Daisyworld," *Tellus* 35B (1983): 286–89.
7. James E. Lovelock, *The Vanishing Face of Gaia: A Final Warning* (London: Penguin Books, 2010), 115.

8. James E. Lovelock, *Gaia: A New Look at Life on Earth*, rev. ed. (Oxford: Oxford University Press, 2000), 113.
9. Lovelock, *Homage to Gaia*, 264.
10. Lovelock, *Gaia: A New Look at Life on Earth*, xi.
11. Lovelock, *The Ages of Gaia*, 11.
12 Lovelock, *Gaia: A New Look at Life on Earth*, viii.
13 Lovelock, *Gaia: A New Look at Life on Earth*, xii–xiii.
14 Stephen J. Gould, "Kropotkin Was No Crackpot," *Natural History* 106 (1997): 12–21.
15 Lovelock, *Homage to Gaia*, 241–79.

MODULE 8
PLACE IN THE AUTHOR'S WORK

KEY POINTS

- The Gaia hypothesis* has been the main focus of Lovelock's work for half a century. The idea of self-regulating Earth* was first presented at an academic conference in the mid–1960s.
- After *Gaia* was published, Lovelock and some collaborators began a decade-long effort to develop the concept into a broader Gaia theory* that would be testable and more scientifically thorough.
- Because Gaia did not fit entirely into any existing academic discipline, the new academic field of Earth system science* emerged.

Positioning

James E. Lovelock, the author of *Gaia: A New Look at Life on Earth*, has always cherished the freedom to follow his own ideas; for this reason, he never associated himself for long with a particular research or academic organization. This is also a reflection of his view that scientific inquiry is a form of direct engagement with the world.[1]

As an independent scientist and inventor, Lovelock invented a number of scientific devices, with the electron capture detector*—a device used to detect and measure atmospheric gases such as chlorofluorocarbons* ("CFCs")—being the most important. Lovelock writes that "electron capture detectors were undoubtedly the most valued of the trade goods which enabled me to pursue my quest for Gaia through the various scientific disciplines, and indeed to travel literally around the Earth itself."[2] Although this invention has nothing to do with the Gaia hypothesis, it gave him access to NASA's* space research laboratory, where the initial thoughts of this hypothesis* were seeded, later bringing him fame and recognition worldwide.

After publication of the first Gaia book in 1979, Lovelock's career took a slightly different turn. Other than concentrating on scientific invention, he became much more engaged in research on Gaia and in writing books and articles in scientific journals. The main focus of his research was further development of his Gaia hypothesis. Over the course of more than a decade, he consulted scientists from several disciplines as he sought to expand the Gaia hypothesis beyond the relatively simple idea of a self-regulating Earth. With the help of collaborators he developed the hypothesis into a theory* able to explain how human interventions in the environment such as the emission of greenhouse gases* like carbon dioxide* and methane* would interfere with the Earth's homeostatic* condition, endangering its delicate climate system.[3] "Homeostatic" here refers to an inclination toward balance, according to which living things hold themselves despite environmental change.

Throughout his career, Lovelock has written more than 200 scientific articles and published 10 books on Gaia. *Gaia: A New Look at Life on Earth*, his first book, remains his most famous.

"As a scientist, I submit wholly to scientific discipline and this is why I sanitized my second book, The Ages of *Gaia, and hopefully made it acceptable to scientists. As a man, I also live in the gentler world of natural history, where ideas are expressed poetically and so that anyone interested can understand, and that is why this book [Gaia] remains almost unchanged."*

—James E. Lovelock, *Gaia: A New Look at Life on Earth*

Integration

Lovelock spent the first part of his professional career in medical research.[4] He conceived the main ideas of a self-regulating Earth at the age of 45 while working at NASA;* these developed into the Gaia hypothesis in the mid-1960s. Since then, the hypothesis has been Lovelock's primary area of research and publication. In his mid-90s at the time of writing, Lovelock is still active in Gaian research. His latest book on the subject, *A Rough Ride to the Future*, was published in 2015.[5]

Lovelock writes "the quest for Gaia has been a battle all the way."[6] But although the hypothesis has been fiercely debated in various scientific circles, even its most persistent critic, the US Earth scientist* James Kirchner,* noted that it has prompted many other scientists to form hypotheses of their own.[7] (Earth science is a discipline drawing on fields as diverse as geology,* chemistry,* biology,* and climatology, conducted to understand our planet's deep history and functioning as a system). Indeed, through collaboration with several scientists, in particular Lynn Margulis* and Andrew Watson,* the hypothesis evolved into a testable theory.

The value of any scientific theory is judged by the accuracy of its predictions.[8] Lovelock's Gaia theory made 10, among which were the lifelessness of Mars, and the critical roles of microorganisms and biodiversity* in the regulation of the planet's climate. So far, eight of these predictions have been confirmed or, at least, become generally accepted by scientists.[9] In his book *On Gaia* (2013), the environmental scientist and Gaian critic Toby Tyrrell* writes that "Lovelock was correct [that life has altered the Earth] both in terms of biological control over aspects of seawater chemistry, and also in terms of the composition of gases in the atmosphere."[10]

The Gaia theory, which draws on many academic disciplines—from astrobiology* (the branch of science concerned with life beyond Earth) to ecology*—has also made significant contributions to scientific research and discovery in many fields such as astrophysics* (the branch of astronomy concerned with the study of the physical characteristics of stars, galaxies, planets, and so on), biology, and Earth science. Finally, the concept and applications of Gaia theory have led to the development of an entirely new, interdisciplinary academic field, Earth system science.[11]

Significance

Lovelock's Gaia hypothesis has influenced many academics, scientists, and politicians over last 40 years. The former Czech president Vaclav Havel said, for example, that "according to the Gaia hypothesis, we are parts of a greater whole. Our destiny is not dependent merely on what we do for ourselves but also on what we do for Gaia as a whole."[12] But the hypothesis was not easily accepted by the scientific community.

Although Lovelock is well known for his scientific inventions in his field, it was the Gaia hypothesis that made him world famous. *Gaia: A New Look at Life on Earth* was bold and radical enough to spark a debate in the scientific community that has continued for decades—demonstrating the significance of the work in contemporary science.

The scientific database ScienceDirect* returns 146 books and 373 scientific articles in response to a search for the phrase "Gaia hypothesis." Scientific articles are still being published on the topic today. In 2006, the world's oldest geological professional body, the Geological Society of London, awarded Lovelock the Wollaston medal,* its highest honor, for his lifelong contribution to the study of the Earth.[13]

---

1. John Gray, "James Lovelock: A Man for All Seasons," *New Statesman*, March 27, 2013, accessed December 21, 2013, http://www.newstatesman.com/ culture/culture/2013/03/james-lovelock-man-all-seasons.
2. See James E. Lovelock, "Preface," in *Gaia: A New Look at Life on Earth*, by James E. Lovelock, rev. ed. (Oxford: Oxford University Press, 2000), xvii.
3. James E. Lovelock, *The Vanishing Face of Gaia: A Final Warning* (London: Penguin Books, 2010), 23–45.
4. James E. Lovelock, *Homage to Gaia: The Life of an Independent Scientist*, rev. ed. (London: Souvenir Press Ltd., 2014), 69–104.
5. James E. Lovelock, *A Rough Ride to the Future* (London: Penguin Books, 2015).
6. Lovelock, *Homage to Gaia*, 278.

7. James W. Kirchner, "The Gaia Hypothesis: Conjectures and Refutations," *Climatic Change* 58 (2003): 21.
8. Lovelock, *The Vanishing Face of Gaia*, 116–17.
9. Lovelock, *The Vanishing Face of Gaia*, 116.
10. Toby Tyrrell, *On Gaia: A Critical Investigation of the Relationship between Life and Earth* (Princeton: Princeton University Press, 2013), 202.
11. James E. Lovelock, "Geophysiology, the Science of Gaia," *Reviews of Geophysics* 27 (1989): 215–22.
12. Lovelock, *Gaia: A New Look at Life on Earth*, x.
13. Lovelock, *Rough Ride*, 77. This reference gives the wrong year (2003) for the award. The award was given in 2006. See "Wollaston Medal Citation," James Lovelock official website, accessed March 4, 2016, http://www. jameslovelock.org/page7.html.

SECTION 3
IMPACT

MODULE 9

THE FIRST RESPONSES

KEY POINTS

- The most important criticism was that Lovelock's Gaia hypothesis* was "scientifically untestable," simply teleological* (founded on a belief that nature has a purpose), and goal-seeking* (based on repeated analyses looking for a particular answer).
- Lovelock responded to criticisms of the Gaia hypothesis by writing his second book, *The Ages of Gaia*—written specifically for scientists.
- SWith evidence that organisms could indeed regulate their own environments, serving to change the status of the hypothesis to that of a testable theory, Gaian principles moved closer to acceptance by the scientific community.

Criticism

James E. Lovelock's Gaia hypothesis initially met with fierce disapproval. One of the strongest criticisms came from Heinrich Holland,* a geochemist* at Harvard University, who rejected the idea that living organisms controlled the regulation of Earth's atmosphere and its climate (geochemistry is the study of the chemical composition of, and chemical changes in, the solid matter of the Earth). Holland insisted climate was controlled by geochemical and geophysical* processes, and no life was actively involved.[1]

Damaging criticism also came from an American biologist, Ford Doolittle,* who reviewed the Gaia hypothesis in the journal *CoEvolution Quarterly* in 1981. Doolittle argued that there was no evidence that individual organisms could provide a cybernetic feedback mechanism*—a physical, biological, or social response within a system that influences the continued activity or productivity of that system—as the hypothesis proposes.[2] He concluded that the Gaia hypothesis was an unscientific theory without any explained mechanism.

Another evolutionary biologist, Britain's Richard Dawkins,* criticized the Gaia hypothesis in *The Extended Phenotype: The Gene as the Unit of Selection* (1982) by arguing that organisms could not act as a unified group because this would require foresight and planning.[3] Echoing Doolittle's comments, he also rejected the idea that a cybernetic feedback mechanism could stabilize the system. Dawkins insisted that there was no way for evolution by natural selection to lead to altruism* (selfless behavior) on a global scale, as was proposed by Lovelock.[4] Dawkins argued that the Gaia hypothesis was teleological and goal-seeking.

The American evolutionary biologist Stephen J. Gould* described the Gaia hypothesis as a metaphorical* description of Earth's processes—that is, a figure of speech not to be taken literally;[5] in 1989, meanwhile, the Earth scientist* James Kirchner* wrote that "Gaia, in its different guises, is a mixture of fact, theory, metaphor, and wishful thinking."[6] For him, the Gaia hypothesis was not scientifically testable and was rather misleading in its suggestion that the Earth's environmental conditions have somehow been altered to meet the needs of the organisms.[7] The normal understanding of natural selection,* one of the fundamental principles of evolutionary theory, is that it is a process that allows organisms to become better adapted to their changing environments.

Despite Lovelock's declaration that the book was meant for a general audience, scientists criticized Lovelock for not using scientific language.[8] Others said the book was a mere expression of a religious faith or modern spiritual fantasy.[9]

> *"The critics took their science earnestly and to them mere association with myth and storytelling made it bad science ... I have tried ... [to convince scientists] both by rewriting my second book,* The Ages of *Gaia, so that is specifically for scientists, and leaving this book [Gaia] as it was."*
>
> ——James E. Lovelock, *Gaia: A New Look at Life on Earth*

Responses

Lovelock responded to criticisms from fellow scientists in various ways, saying critics did not understand, or that they were judging *Gaia* by the wrong standards. He responded very actively to one particular criticism, that the Gaia hypothesis was not testable. Lovelock's response was to launch a decade-long effort to develop his hypothesis into a scientifically rigorous theory that might become more widely accepted. The result was the Gaia theory.*

In 1983, Lovelock and his then postgraduate student, Andrew Watson,* developed an experimental computer model called "Daisyworld"*[10]—an imaginary Earthlike planet with a sun and a simple ecosystem* of two daisy species, black and white, which compete for space as they grow (an ecosystem can be understood, roughly, as a habitat and the organisms that survive in that habitat). The temperature of the planet is influenced by the proportion of ground covered by the two types of daisy. As the brightness of the sun increases over time, as has happened in reality with our own Sun, the planet becomes increasingly covered by white daisies. These reflect back more of the sun's warmth than the black daisies, and hence keep the planet cool.[11]

The model supported the idea that the temperature of an environment could be regulated by two competing species of plants living in the same environment so that ultimately an optimal temperature is reached. The Daisyworld model was accepted by many scientists, and particularly by mathematicians. One supporter was the mathematician Peter Saunders,* who commented that the model was worthy of study. Another supporter, the British Earth system scientist* Timothy Lenton,* has published a number of scientific papers on the implications of the Gaia theory and its mathematical basis.

In addition to developing the Gaia theory, Lovelock wrote a number of books on Gaia, including *The Ages of Gaia: A Biography of our Living Earth* (1988),[12] in which he explained his ideas using scientific language and responded to criticisms about his first book. Many scientific meetings and conferences took place to discuss the Gaia theory. For example, in 1994, scholars from many different disciplines discussed different aspects of Gaia theory and Earth's regulatory processes at a meeting titled "The Self-Regulating Earth."[13]

Conflict and Consensus

Lovelock's Gaia theory has become more complex and open to modification over the last four decades. Gaian thinking evolved from the provocative hypothesis that life controls planetary conditions for its own benefit, to a more robust and sophisticated theory that positions life as a key player in shaping the planet Earth.[14]

The concept of a cybernetic feedback mechanism for the earth's atmospheric regulation has been relevant to the debate around global warming,* as highlighted in 2001 by the Amsterdam Declaration on Global Change.* The environmental dimensions of Gaia theory revolve around two main concepts: the consequences of human-driven disruptions of the biosphere,* and the implications of habitat destruction and fragmentation of the Earth's ecosystems. While small-scale disturbances can be absorbed by the biosphere, large-scale disruptions sooner or later trigger far-reaching and uncontrollable consequences in the global climate.[15]

Gaian thought has also been addressed by sociologists (those studying society and social behavior). The sociologist Eileen Crist* and the ecologist* H. Bruce Rinker* have written:"The anthropogenic* amplification of the greenhouse effect* underway is rapid and large enough that it may unleash positive feedback ... positive feedback, in turn, can trigger runaway heating. Such an eventuality will not only cause widespread human suffering, it will transform the Earth into a biological wasteland."[16]

In that sense, the Gaia hypothesis can be seen as a "game

changer" because it has essentially shifted people's perception toward a more sustainable Earth.

While no consensus has been reached on whether the planet Earth is a self-regulating, living entity, Lovelock's Gaia and its central ideas, particularly the role of microorganisms in regulating Earth's atmosphere, are widely discussed in scientific work. Lovelock writes in his autobiography "if they [scientists] must reject Gaia as the name of their new science I hope that they will choose 'Earth System Science' as a sensible alternative."[17]

---

1. James E. Lovelock, *The Vanishing Face of Gaia: A Final Warning* (London: Penguin Books, 2010), 111–12.
2. W. Ford Doolittle, "Is Nature Really Motherly?," *CoEvolution Quarterly* 29 (1981): 60.
3. See Richard Dawkins, *The Extended Phenotype: The Gene as the Unit of Selection* (Oxford: Oxford University Press, 1982), 1–307.
4. See Dawkins, *The Extended Phenotype*, 234–36.
5. Stephen J. Gould, "Kropotkin Was No Crackpot," *Natural History* 106 (1997): 12–21.
6. James W. Kirchner, "The Gaia Hypothesis: Fact, Theory, and Wishful Thinking," *Climatic Change* 52 (2002): 391.
7. James W. Kirchner, "Gaia Hypothesis: Can It Be Tested?," *Reviews of Geophysics* 27, no. 2 (1989): 223.
8. See James E. Lovelock, "Preface," in *Gaia: A New Look at Life on Earth*, by James E. Lovelock, rev. ed. (Oxford: Oxford University Press, 2000), x–xi.
9. Eileen Crist and H. Bruce Rinker, "One Grand Organic Whole," *Gaia in Turmoil: Climate Change, Biodepletion, and Earth Ethics in an Age of Crisis* (Cambridge: The MIT Press, 2010), 7.
10. Andrew J. Watson and James E. Lovelock, "Biological Homeostasis of the Global Environment: The Parable of Daisyworld," *Tellus* 35B (1983): 286–89.
11. Toby Tyrrell, *On Gaia: A Critical Investigation of the Relationship between Life and Earth* (Princeton: Princeton University Press, 2013), 25.
12. James E. Lovelock, *The Ages of Gaia: A Biography of Our Living Earth*, rev. ed. (Oxford: Oxford University Press, 1995).

13. James E. Lovelock, *Homage to Gaia: The Life of an Independent Scientist*, rev. ed. (London: Souvenir Press Ltd., 2014), 274–75.
14. Eileen Crist and H. Bruce Rinker, "One Grand Organic Whole," in *Gaia in Turmoil: Climate Change, Biodepletion, and Earth Ethics in an Age of Crisis*, ed. Eileen Crist and H. Bruce Rinker (Cambridge: The MIT Press, 2010), 8.
15. Crist and Rinker, "One Grand Organic Whole," 11–12.
16. Crist and Rinker, "One Grand Organic Whole," 14.
17. Lovelock, *Homage to Gaia*, 278–79.

MODULE 10
THE EVOLVING DEBATE

KEY POINTS

- The Gaia hypothesis has contributed significantly to the field of climate science,* particularly in the deep understanding of global climate and the role of microorganisms in regulating the Earth's atmosphere.
- Over the years, the concept of a self-regulating Earth* has slowly come to influence contemporary science by inspiring scientists, researchers, students, and politicians.
- The impact of the Gaia hypothesis on academic research and policies has been widespread, including the emergence of the new discipline of Earth system science,* and the relevance to the current debate about anthropogenic* climate change* through global warming.*

Uses and Problems

James E. Lovelock has continued to write about Gaia since the publication of his *Gaia: A New Look at Life on Earth* in 1979. In his *The Vanishing Face of Gaia* (2010) he describes how from the early 1980s to the mid-1990s it was impossible to publish any paper on the subject in any mainstream journal.[1] Since then, however, Lovelock's controversial hypothesis has influenced academics, scientists, politicians, and the general public for many years. In a critical book about the Gaia hypothesis in 2013, the British Earth system scientist Toby Tyrrell* writes:"In the thirty years or so since its inception, this Gaia hypothesis has vigorously inspired, infuriated, and intrigued a whole generation of environmental scientists."[2] From the mid-1960s to the late 1980s, Lovelock's hypothesis was severely criticized by geologists,* evolutionary biologists,* and planetary scientists.* Lovelock took critics' comments on board, and transformed his hypothesis into the more sophisticated, scientifically testable Gaia theory; in an updated edition of *Gaia*, he writes:"Gaia hypothesis was a vague speculation before the blood was drawn to leave the ... more scientifically acceptable Gaia theory. For this I am grateful to the critics."[3]

Lovelock's first book, *Gaia*, did not convince many scientists—but a few academics welcomed the ideas it presented. One of these was René Dubos,* a French American microbiologist, who wrote in a review article that he read it with "immense pleasure,"[4] accepting that the Earth without life would produce an atmosphere filled with carbon dioxide gas that cannot support life.

In 1988, Lovelock published his second book, *The Ages of Gaia*—a more technical version of the first book. *The Ages of Gaia* laid out the main ideas more scientifically, explaining, for example, the critical role of ocean algae* (a plantlike organism) in climate regulation. In 1987, Lovelock and three collaborators, led by the American climate scientist Robert Charlson,* published a paper on what they called the CLAW hypothesis.*[5] The hypothesis, which draws its name from the collaborator's initials (Charlson, Lovelock, Andreae,* and Warren),* refined the Gaian concept of a cybernetic feedback mechanism* by attributing it specifically to microscopic ocean organisms known as phytoplankton.* A search for the phrase

"CLAW hypothesis" on the database ScienceDirect finds 33 papers and 16 books by numerous authors who have mentioned the subject.

> *"Gaia theory aims to be consistent with evolutionary biology and views the evolution of organisms and their material environment as so closely coupled that they form a single, indivisible, process."*
>
> ——Timothy M. Lenton, "Gaia and Natural Selection," *Nature*

Schools of Thought

In the 1980s, Lovelock and the British marine and atmospheric scientist Andrew Watson* developed the digital model they called Daisyworld* in order to translate the hypothesis into a scientifically defensible theory through testable proof.

Toby Tyrrell writes that "as Lovelock, now in his nineties, has become less active, others have taken up the torch."[6] For instance, Timothy Lenton,* an Earth system scientist at the University of Exeter in England, has published more than 20 articles on Gaia theory since 1997. Lenton is one of Watson's students; in 2011, the two published a book, *Revolutions that Made the Earth*, which expands on the ideas of the Gaia hypothesis.[7] He had previously made a significant contribution in a review article on the Gaia theory published in the journal *Nature* in 1998 in which he argued that the process of natural selection*—as famously described by Charles Darwin*—could be seen as an integral part of Gaia.[8]

Gaian models suggest that we have to think about organisms

and their environments as one whole to understand fully which traits (characteristics that can be passed from generation to generation) come to persist and dominate. Scientists and researchers from around the world recognized in the Amsterdam Declaration on Global Change* of 2001 that "the Earth system behaves as a single, self-regulating system comprised of physical, chemical, biological and human components"—an idea central to Lovelock's work, even if the Declaration does not include the word "Gaia."[9]

In Current Scholarship

Today, scientists from many disciplines recognize Gaia theory as one of the most influential theories in contemporary science. For instance, Gaia theory adds to Darwin's vision that the success of species depends upon coherent coupling between the evolution of the organisms and the evolution of their material environment.[10] Despite his persistent criticism of the hypothesis, the Earth scientist James Kirchner* has nevertheless described it as "a fruitful hypothesis generator"—an inspiration for other research—for its having "prompted many intriguing conjectures about how biological processes might contribute to planetary-scale regulation of atmospheric chemistry and climate."[11]

Today, supporters of Lovelock's Gaia hypothesis come from disciplines as diverse as astrobiology* (inquiry into life elsewhere to the Earth), biology,* Earth system science, ecology,* environmental science,* and climate science. With his students, Andrew Watson (codeveloper of the Daisyworld model) has developed other mathematical models to represent regulation of atmospheric composition through geological time—which is to say through very long periods.[12]

---

1. James E. Lovelock, *The Vanishing Face of Gaia: A Final Warning* (London: Penguin Books, 2010), 111.
2. Toby Tyrrell, *On Gaia: A Critical Investigation of the Relationship between Life and Earth* (Princeton: Princeton University Press, 2013), ix.
3. See James E. Lovelock, "Preface," in *Gaia: A New Look at Life on Earth*, by James E. Lovelock, rev. ed. (Oxford: Oxford University Press, 2000), xv.
4. René Dubos, "Gaia and Creative Evolution," *Nature* 282 (1979): 154–55.
5. Greg Ayers and Jill Cainey, "The CLAW Hypothesis: a Review of the Major Developments," *Environmental Chemistry* 4 (2007): 366–74.
6. Tyrrell, *On Gaia*, 3.
7. Timothy M. Lenton and Andrew Watson, *Revolutions that Made the Earth* (Oxford: Oxford University Press, 2013).
8. Timothy M. Lenton, "Gaia and Natural Selection," *Nature* 394 (1998): 447.
9. James E. Lovelock, "The Living Earth," *Nature* 426 (2003): 769–70.
10. James E. Lovelock, "Geophysiology, the Science of Gaia," *Reviews of Geophysics* 27 (1989): 222.
11. James W. Kirchner, "The Gaia Hypothesis: Conjectures and Refutations," *Climatic Change* 58 (2003): 21.
12. James E. Lovelock, *The Vanishing Face of Gaia: A Final Warning* (London: Penguin Books, 2010), 105–22.

MODULE 11

IMPACT AND INFLUENCE TODAY

KEY POINTS

- Lovelock's Gaia hypothesis* is still a hot topic for scientific debate and discovery nearly 40 years after the book's publication.
- Some critics of the Gaia theory* say that its claim that living things have a measure of control over the environment cannot be squared with the evolutionary theory derived from the work of Charles Darwin,* according to which living things adapt to the environment.
- Other critics challenge the predictive power of Gaia theory, and suggest that life may actually destroy the planet, rather than save it.

Position

James E. Lovelock's radical Gaia hypothesis as set out in his *Gaia: A New Look at Life on Earth* has continued to live among contemporary scientists and academics as a provocative scientific topic. Even after 40 years, the hypothesis is still being reexamined and reinterpreted by scientists of various fields.[1]

The Gaia theory is highly interdisciplinary, encompassing many disciplines ranging from astrobiology* to ecology.* Over a period of three decades, this theory has made significant contributions to many pieces of scientific research and to discoveries in astrophysics,* biology,* Earth science* and ecology. The concept behind Gaia theory has led to the development of a new multidisciplinary subject, geophysiology,* also known as Earth system science,* which treats the Earth as an interlinked system and tries to get a better understanding of the physical, chemical, biological, and human interactions that determine the planet's past, current, and future states.

Another noteworthy contribution Gaia theory has made to climate science* and environmental science* is that it helps us better understand the current debate about global warming* and climate change.* The discipline of ecological philosophy, also known as deep ecology,* a way of looking at world problems that unites thinking, feeling, spirituality, and action, has been very much influenced by the Gaia hypothesis. Similarly, the idea of eco-spirituality,* which connects the science of ecology with spirituality, has been inspired by Lovelock's work.

According to the contemporary critic of the Gaia hypothesis Toby Tyrrell*—a professor of Earth system science at the University of Southampton—the Gaia hypothesis has achieved a degree of scientific respectability.[2] But while it is now accepted gladly by some, it also continues to stimulate intense debate.[3] Another persistent critic of the hypothesis, the Earth scientist* James Kirchner,* has written research articles[4,5] in scientific journals disputing the theory's predictive power. The importance of the Gaia hypothesis in explaining several vital mechanisms operating at the planetary scale on the Earth (such as the regulation of atmospheric gases) has guaranteed its continuing relevance in modern science.[6]

> *"Gaia theory proposes that organisms inflicting damage on their surroundings will eventually reap harsh consequences when feedback comes back to haunt them. We are currently experiencing such feedback in the form of climate change, ozone depletion, endocrine disruption, and desertification."*
>
> ——Eileen Crist and H. Bruce Rinker, "One Grand Organic Whole," *Gaia in Turmoil*

Interaction

Scientists from disciplines such as geology,* evolutionary biology,* and planetary science* have expressed serious reservations about the idea of a self-regulating, living planet. The main challenge is that the Gaia hypothesis suggests biotic control of the Earth's environment, meaning control by the biota*—the plant and animal life in the environment. This conflicts with evolutionary theory, according to which organisms adapt to their own environments.[7] Charles Darwin's theory of evolution through natural selection* holds that organisms with traits favoring survival will tend to live to reproduce and pass on those qualities to their offspring, while those that do not survive to reproduce will become extinct.[8]

Tyrrell and others have argued that natural selection operates according to the simple rule that whatever works best in time and space will be favored, regardless of future implications for the wider ecosystem or global impact.[9] Furthermore, global biota are not a closely related family group, and it is not possible that cooperation takes place at a scale as large as the whole Earth.[10]

The Continuing Debate

In his 2013 book *On Gaia: A Critical Investigation of the Relationship between Life and Earth*, Tyrrell considered why the Gaia hypothesis has an enduring appeal for some scientists and members of the general public. First, he thinks that Gaia is big-picture science that offers answers to deep questions such as that of why the Earth has remained continuously habitable for so long.[11]

Second, it suggests mechanisms for how our planet can cope in the future as it continues to be affected by global warming. Both James Kirchner and Toby Tyrrell argue that in order to protect planet Earth as a life-support system, our understanding of its natural processes must be based on a correct view. In this context, Kirchner provides some examples that challenge the predictive power of Gaia theory. For instance, Gaia theory predicts that biological processes should tightly regulate the makeup of the Earth's atmosphere, but rates of carbon uptake by microbes have increased by only about 2 percent in response to a 35 percent rise in atmospheric carbon dioxide* gas since preindustrial times.[12] Lovelock argues that Earth's system of self-regulation is now being overwhelmed by anthropogenic* greenhouse gas* pollution.

Another example of how the Gaia hypothesis has affected the contemporary intellectual world is the development of an anti-Gaian concept known as the Medea hypothesis* by Peter Ward,* an American paleontologist*—a researcher in fossils—at the University of Adelaide in Australia.[13] Like the Gaia hypothesis, the Medea hypothesis is named after a character from ancient Greek mythology: in this case, Medea—the wife of the mythological hero Jason. Ward argues that life is self-destructive, and gives several examples of mass extinction* in Earth's history. Ward's argument—which is essentially the opposite of the Gaia hypothesis—is that life will cause its own end by warming the biosphere about 1 billion years from now.[14]

---

1. Toby Tyrrell, *On Gaia: A Critical Investigation of the Relationship between Life and Earth* (Princeton: Princeton University Press, 2013), 1–6.
2. Tyrrell, *On Gaia*, 1–6.
3. Tyrrell, *On Gaia*, 3.
4. James W. Kirchner, "Gaia Hypothesis: Can It Be Tested?," *Reviews of Geophysics* 27, no. 2 (1989): 223.
5. James W. Kirchner, "The Gaia Hypothesis: Conjectures and Refutations," *Climatic Change* 58 (2003): 21–45.
6. Crispin Tickell, "Scientists on Gaia," *Financial Times* 2002, accessed December 23, 2013, http://www.crispintickell.com/page19.html.
7. Timothy M. Lenton, "Gaia and Natural Selection," *Nature* 394 (1998): 439–47.
8. Charles Darwin, *On the Origin of Species by Means of Natural Selection, or the Preservation of Favoured Races in the Struggle for Life* (London: John Murray, 1859).
9. Tyrrell, *On Gaia*, 34.
10. Tyrrell, *On Gaia*, 40.
11. Tyrrell, *On Gaia*, 1–6.
12. Kirchner, "The Gaia Hypothesis: Conjectures and Refutations," 21.
13. Peter Ward, *The Medea Hypothesis: Is Life on Earth Ultimately Self-Destructive?* (Princeton: Princeton University Press, 2009), 208.
14. Moises Velasquez-Manoff, "The Medea Hypothesis: A Response to the Gaia Hypothesis," *The Christian Science Monitor*, February 12, 2010, accessed December 28, 2013, http://www.csmonitor.com/Environment/Bright-Green/2010/0212/The-Medea-Hypothesis-A-response-to-the-Gaia-hypothesis.

MODULE 12
WHERE NEXT?

KEY POINTS

- The Gaia hypothesis* has influenced scientific theories and discoveries over the last four decades and is relevant to academic research in the subject of global climate change.*
- Today, climate scientists modeling Earth's future climate are considering the suggestions made by the Gaia hypothesis—an important impact of Lovelock's groundbreaking work.
- The ideas offered by *Gaia: A New Look at Life on Earth* have led to the development of an entirely new discipline: Earth system science.*

Potential

Thanks to his publications such as *Gaia: A New Look at Life on Earth*, James E. Lovelock has been recognized by popular American magazine *Rolling Stone* as one of the twentieth century's most influential scientists.[1] According to the text, written for a nonscientific audience, living organisms* and their physical environment form a complete entity that controls the Earth's atmosphere and climate. The idea of a self-regulating Earth* was so radical at that time that many scientists criticized the Gaia hypothesis as simply bad science. In 1988, Lovelock published his second book, *The Ages of Gaia*, specifically for scientists.

As the Gaia theory* started to receive attention not only from the general public but from politicians, academics, and scientists, Lovelock wrote five more books about the Gaia hypothesis.[2]

According to a piece in the British *Daily Telegraph* newspaper, the book "could prove to be one of the twentieth century's most important pieces of polemic" ("polemic" here refers to a strongly worded work written with the intention to persuade).[3]

"Today [the Gaia hypothesis] is probably more widely credited than ever," writes Toby Tyrrell* in his critical 2013 book *On Gaia*.[4] The hypothesis has the potential to affect the debate over the Earth's climate and how it is regulated. In *Gaia* and subsequent books, Lovelock emphasizes anthropogenic* (human-caused) global warming.* Increased concentration of greenhouse gases* (particularly carbon dioxide*) has been driven by industrialization and increased burning of oil, gas, and coal. As the greenhouse effect* increases, the atmosphere traps more heat, and the loss of light-reflecting surfaces, mainly ice and snow, means that more light (and heat) is absorbed and less reflected, threatening to create a positive feedback effect and trigger runaway heating.[5]

Under these circumstances, the Gaia hypothesis can offer scientists and policymakers a unique perspective from which current environmental problems can be viewed. Indeed, the Gaia hypothesis has already made significant contributions in understanding the issue of global warming.[6]

> *"We need the people of the world to sense the real and present danger so that they will spontaneously mobilize and unstintingly bring about an orderly and sustainable withdrawal to a world where we try to live in harmony with Gaia."*
>
> ——James E. Lovelock, *The Revenge of Gaia*

Future Directions

Lovelock writes in his *The Vanishing Face of Gaia: A Final Warning* (2010) that changes in the Earth's climate could lead to the disappearance of sensitive ecosystems,* and potentially endanger the existence of human life.[7] Lovelock strongly criticizes some climate scientists and politicians for failing to see the Earth as a self-regulating, living entity.[8] For a long time, they considered the Earth as a solid rock and did not consider biotic* (biological) interactions with the Earth's physical environments.[9] Recently, the Intergovernmental Panel on Climate Change (IPCC)* has considered new climate models, known as Earth system models,* that recognize the interaction of multiple factors in affecting the climate.

According to an article in the British *Guardian* newspaper, Lovelock "is not a doom-monger but a practical problem-solving man, with suggestions for alleviating the climate crisis at many levels."[10] Lovelock strongly believes that the concept of Gaia will develop further in the future. The Earth system scientist Timothy Lenton,* for instance, has been leading research on the Gaia hypothesis. In his recent book *The Vanishing Face of Gaia*, Lovelock claims that new technologies such as geoengineering*—deliberate large-scale intervention in the Earth's natural systems—will emerge to counteract climate change.

Summary

Many applications of the Gaia hypothesis have materialized since the publication of *Gaia: A New Look at Life on Earth* in 1979. Lovelock provided a number of interesting ideas in his book for further academic research. His Gaia theory has prompted some intriguing interpretations of the way biological processes could contribute to planetary-scale regulation of atmospheric chemistry and climate.[11] In *Gaia*, Lovelock argues that Earth is a self-regulating system. Several planetary mechanisms such as the regulation of atmospheric gases, atmospheric temperature, and the salt levels of oceans are explained in the book. Lovelock also explains how these systems are controlled by living organisms on Earth through a cybernetic feedback mechanism*—an automatic control system that makes adjustments in response to change. All these ideas described in Lovelock's *Gaia* have significantly contributed to academic research and applications in the field of ecology* and climate science* over the last 40 years.

The Gaia hypothesis is interdisciplinary in nature, drawing on the aims and knowledge of many different academic disciplines; it does not fall within a single one. Lovelock proposed the study of Earth systems within a new discipline, geophysiology,* or Earth system science.[12] Gaia's ideas were also considered by academics of other disciplines such as ecology, marine biology,* and climate science. One of the most remarkable applications of the Gaia hypothesis was the development of a mathematical model called Daisyworld*[13]—now used by ecologists to test the role of biodiversity* and stability of ecosystems for a healthy living environment.

A more recent application of the Gaia hypothesis is seen in school education.[14] A 2009 study in Brazil has found that the Gaia hypothesis can contribute to the understanding of human activities and contemporary environmental issues such as global warming and climate change in science education at school. The study has also found that the interdisciplinary nature of Lovelock's controversial Gaia hypothesis makes an interesting and effective tool for cross-disciplinary learning at school.

---

1. Jeff Goodell, "James Lovelock, the Prophet," *Rolling Stone*, November 1, 2007, accessed December 23, 2013, http://www.rollingstone.com/politics/ news/james-lovelock-the-prophet–20071101.
2. James E. Lovelock, *Homage to Gaia: The Life of an Independent Scientist* (2000), *Healing Gaia: The Practical Science of Planetary Medicine* (2001), *Gaia: Medicine for an Ailing Planet* (2005), *The Revenge of Gaia: Why the Earth is Fighting Back—and How we Can Still Save Humanity* (2007), and *The Vanishing Face of Gaia: A Final Warning* (2010).
3. James Flint, "Earth—The Final Conflict," The *Daily Telegraph*, February 6, 2006, accessed December 23, 2013, http://www.telegraph.co.uk/culture/ books/3649909/Earth-the-final-conflict.html.
4. Toby Tyrrell, *On Gaia: A Critical Investigation of the Relationship Between Life and Earth* (Princeton: Princeton University Press, 2013), ix.
5. Eileen Crist and H. Bruce Rinker, "One Grand Organic Whole," in *Gaia in Turmoil*, ed. Eileen Crist and H. Bruce Rinker (Cambridge: The MIT Press, 2010), 1–20.
6. James E. Lovelock, *A Rough Ride to the Future* (London: Penguin Books, 2015), 85–103.
7. James E. Lovelock, *The Vanishing Face of Gaia: A Final Warning* (London: Penguin Books, 2010), 1–45.
8. James E. Lovelock, *The Revenge of Gaia: Why the Earth is Fighting Back—and How We Can Still Save Humanity* (London: Penguin Books, 2007), 61–83.
9. Lovelock, *Vanishing Face of Gaia*, 1–45.
10. Peter Forbes, "Jim'll Fix It," The *Guardian*, February 21, 2009, accessed December 23, 2013, http://www.theguardian.com/culture/2009/feb/21/ james-lovelock-gaia-book-review.
11. James W. Kirchner, "The Gaia Hypothesis: Conjectures and Refutations," *Climatic Change* 58 (2003): 21.
12. James E. Lovelock, "Geophysiology, the Science of Gaia," *Reviews of Geophysics* 27 (1989): 215–22.
13. Andrew J. Watson and James E. Lovelock, "Biological Homeostasis of the Global Environment: The Parable of Daisyworld," *Tellus* 35B (1983): 286–89.
14. Ricardo Santos do Carmo, Nei Freitas Nunes-Neto, and Charbel Nino El-Hani, "Gaia Theory in Brazilian High School Biology Textbooks," *Science & Education* 18 (2009): 469–501.

GLOSSARY OF TERMS

1. **Algae:** plantlike organisms that can make food in the presence of sunlight by photosynthesis.
2. **Altruism:** a human behavior that displays a desire to help others selflessly.
3. **American Geophysical Union:** a nonprofit organization founded for the advancement and wider understanding of research in geophysics.
4. **Ammonia:** a strong, colorless gas composed of nitrogen and hydrogen with a characteristic pungent odor.
5. **Amsterdam Declaration on Global Change:** a declaration made by the scientific communities of four international global change research programs, recognizing that ever-increasing human modifications of the global environment have great implications for human well-being, in addition to the threat of climate change.
6. **Anthropogenic:** something caused or influenced by humans.
7. **Apollo mission:** a spaceflight program carried out by the National Aeronautics and Space Administration (NASA). Apollo astronauts became the first human beings on the Moon between 1969 and 1972.
8. **Artificial satellite:** a human-built object orbiting the Earth and other planets in the solar system.
9. **Astrobiology:** the science concerning the origin and evolution of life beyond Earth, in the universe.
10. **Astronomer:** a scientist who studies stars, planets, moons, comets, galaxies, and so on.
11. **Astrophysics:** the branch of astronomy that studies the physical characteristics and composition of celestial objects such as stars, moons, planets, and so on.
12. **Atmospheric gases:** gases present in the Earth's atmosphere such as oxygen, carbon dioxide, nitrogen, and so on.
13. **Bacteria:** single-celled, microscopic organisms that are found everywhere both inside and outside of human bodies. Some bacteria are helpful for human health but the majority are harmful.
14. **Biochemist:** a scientist with a qualification in biochemistry—the branch of science that studies chemical processes in living organisms.

15. **Biodiversity:** the diversity (variety) of species in any particular location.
16. **Biology:** the scientific study of living things such as plants and animals.
17. **Biologist:** a scientist who focuses on living organisms, including plants and animals.
18. **Biosphere:** the regions of the surface and atmosphere of the Earth or another planetary body that are occupied by living organisms.
19. **Biota:** living organisms such as animals and plants of a region, habitat, or geological period.
20. **Carbon dioxide:** a colorless, odorless, and incombustible gas that is commonly found in the Earth's atmosphere. A greenhouse gas, it is formed during respiration and the decomposition and combustion of organic matter.
21. **CFCs:** See chlorofluorocarbons.
22. **Chemistry:** the branch of science that studies chemical properties of substances and their reactions.
23. **Chlorofluorocarbons:** also known as CFCs, these are nontoxic and nonflammable chemical compounds containing atoms of carbon, chlorine, and fluorine. They are commonly used as coolants.
24. **CLAW hypothesis:** developed by Lovelock and three other scientists, the CLAW hypothesis proposes that microscopic organisms inhabiting the ocean surface are able to regulate their own population; the name is derived from the initials of the four scientists Robert Charlson, James Lovelock, Meinrat Andreae, and Stephen G. Warren.
25. **Climate change:** a long-term change in the planet's weather patterns such as a shift in average temperatures.
26. **Climate science:** the study of climate—the average condition of day-to-day weather measured over a considerably long period.
27. **Continental drift:** a theory proposed by German scientist Alfred Wegener that the continents drift and move position on Earth's surface over millions of years.
28. **Cosmic radiation:** an emission of cosmic rays made up of energetic, subatomic-

sized particles arriving from outside the Earth's atmosphere.

29. **Cybernetic feedback mechanism:** a response within a mechanical, physical, biological, or social system that influences the continued activity or productivity of that system.

30. **Daisyworld:** in a digital model designed by Lovelock and his student Andrew Watson to provide testable evidence for the Gaia hypothesis, Daisyworld was a cloudless hypothetical Earthlike planet with a negligible atmospheric greenhouse. There were two daisy species on the planet—one black, one white. The ground covered by the black (dark) daisy reflected less light than clear ground; the ground covered by the white (light) daisy reflected more light than the bare ground.

31. **Deep ecology:** an all-inclusive approach that combines feeling, spirituality, thinking, and action in seeking to solve the world's problems. It involves moving beyond the self-centred nature of modern culture toward seeing human beings as part of the Earth.

32. **Disequilibrium:** a loss or lack of equilibrium or stability in which opposing forces or influences are not in balance with each other.

33. **Earth science:** a discipline drawing on fields as diverse as geology, chemistry, biology, and climatology conducted to understand our planet's deep history and functioning as a system.

34. **Earth system models:** models that look at the interactions of atmosphere, ocean, land, ice, and biosphere in order to gauge the state of regional and global climatic conditions under a wide variety of settings.

35. **Earth system science:** an interdisciplinary field of study that identifies the Earth as a unified system and looks for a way to understand the interaction of the environment (physical, chemical, and biological) and human activity to establish the state of the planet in the past, present, and future.

36. **Earth system scientist:** a scientist in the field of Earth system science.

37. **Ecology:** the branch of biology that studies the relationship of organisms to each other and to their physical environments.

38. **Eco-spirituality:** a philosophy based in a fundamental belief in the sacredness

of nature, Earth, and the universe.

39. **Ecosystem:** a biological system made up of organisms found within a specific physical environment that interact both with the environment and also with each other.

40. **Electron capture detector:** a device used to detect trace amounts of chemical compounds in a sample.

41. **Environmental science:** a multidisciplinary science that looks at environmental conditions and their effect on organisms living in that environment.

42. **Equilibrium:** a state in which opposing forces or influences are balanced with each other.

43. **Evaporation:** the process by which a substance in a liquid state changes to a gas due to an increase in temperature or pressure or a combination of both.

44. **Evaporites:** a deposit of minerals, formed by the evaporation of salt water.

45. **Evolution:** a gradual change in the characteristics of a group of animals or plants that takes place over successive generations and explains the development of existing species from their dissimilar ancestors.

46. **Evolutionary biology:** a branch of biology that studies the evolution of organisms, especially molecular and microbial evolution. Other areas involve behavior, genetics, ecology, life histories, and development.

47. **Gaia hypothesis:** an idea that the physical and chemical condition of the surface of the Earth, of the atmosphere, and of the oceans has been and is actively made fit and comfortable by the presence of life itself—according to which, as James Lovelock puts it,"the entire surface of the Earth including life is a self-regulating entity."

48. **Gaia theory:** a theory derived from the Gaia hypothesis and substantially proved by a mathematical model developed by James Lovelock and AndrewWatson ("Daisyworld").

49. **Gaian thinking:** the philosophical position that all organisms on Earth, including humans, interact with each other to shape their living environment and keep it fit and comfortable—the way of thinking holistically inspired by the Gaia hypothesis.

50. **Geochemist:** a scholar of the chemical composition of and chemical changes in

the solid matter of the Earth or other planetary bodies.

51. **Geoengineering:** an intentional, large-scale technological intervention in the Earth's natural systems. Geoengineering is often discussed as a techno-fix for combating global climate change.
52. **Geology:** the study of the origin, history, and structure of the Earth—the science dealing with the solid Earth and its rocks.
53. **Geophysicist:** a scholar of the various gravitational, magnetic, electrical, and seismic phenomena (such as earthquakes) that define a planet.
54. **Geophysiology:** the study of interactions among all living organisms on Earth. Geophysiology is a transdisciplinary environment for studying planetary-scale problems.
55. **Geosciences:** the sciences concerned with the Earth such as geology, geophysics, and geochemistry.

56. **Global warming:** a gradual increase in long-term average temperature of the Earth's atmosphere.
57. **Goal-seeking:** the process of calculating an output by performing various "what-if" analyses on a given set of inputs.
58. **Greek mythology:** myths and tales about the ancient Greeks. Greek mythology is concerned with mythological gods and heroes, various other mythological creatures, and the origins and importance of the ancient Greeks' cult and ritual practices.
59. **Greenhouse effect:** the natural process by which the Earth's atmosphere traps some of the Sun's energy, keeping the atmosphere warm enough to support life. It is referred to as the greenhouse effect because a greenhouse works in much the same way, trapping heat within itself.
60. **Greenhouse gases:** gases that contribute to the greenhouse effect by absorbing infrared radiation invisible to human eyes.
61. **Homeostasis or homeostatic condition:** a state of constancy in which living things hold themselves when the environments in which they reside are changing.
62. **Humanism:** a philosophical position that emphasizes human concerns; in its rational

form, it serves as an alternative ethical system to those derived from religion.

63. **Hypothesis:** an idea or concept that is not proven scientifically but leads to further study or discussion. In science, a hypothesis needs to go through much testing before it can earn the status of a theory. In the nonscientific world, the words hypothesis and theory are often used interchangeably.

64. ***Icarus:*** a scientific journal dedicated to the field of planetary science.

65. **Intergovernmental Panel on Climate Change (IPCC):** a scientific intergovernmental body represented by all nations under the auspices of the United Nations.

66. **Marine biology:** the study of marine species or organisms that live in the ocean and other salt-water environments.

67. **Mass extinction:** a relatively sudden decrease in the diversity of animals or plants on a global scale. Throughout the history of life on Earth there have been mass extinctions and they generally occur in a short amount of geological time.

68. **Medea hypothesis:** the idea that life is its own enemy and that nearly all mass extinctions on Earth were caused by life itself. This hypothesis, put forward by American paleontologist Peter Ward, stands in complete contrast to Lovelock's Gaia hypothesis, which suggests that life itself sustains a comfortable living condition on Earth. Gaia views the Earth as a "good mother," whereas Medea views the Earth as an "evil mother."

69. **Metaphor:** a word or phrase generally used to compare two objects, ideas, thoughts, or feelings that are different.

70. **Methane:** a colorless, odorless, and flammable gas. The major constituent of natural gas, methane is the simplest of the hydrocarbons. It is released during the decomposition of plant or other organic compounds. Methane is also considered a greenhouse gas.

71. **NASA:** the National Aeronautics and Space Administration. NASA is the United States government agency responsible for the civilian space program.

72. **National Institute for Medical Research:** a medical research institute situated near London, England. It was set up in 1913 by the UK's Medical Research Council.

73. **Natural selection:** the process through which evolutionary changes occur when individual organisms such as plants and animals that have certain characteristics achieve a greater survival or reproductive rate than other individuals in the same population. The more successful individuals then pass these useful inheritable characteristics on to their offspring.

74. **Organism:** an animal, plant, fungus, or bacterium that is a living biological entity.

75. **Paleontology:** the scientific study of plant and animal fossils.

76. **Photosynthesis:** the process by which green plants and other organisms use light energy (usually from sunlight) to create nutrients from carbon dioxide and water. Plants produce food by this process; oxygen is also released as a waste product.

77. **Physics:** the branch of fundamental science that studies the nature and properties of matter and energy.

78. **Phytoplankton:** tiny marine plants that, in a balanced ecosystem, provide food for a huge range of sea creatures including jellyfish, whales, and snails.

79. **Planetary science:** the study of planets including the Earth, moons, and planetary systems, particularly those belonging to the solar system.

80. **Planetary scientist:** a scholar of planets.

81. **Radiation:** energy transmitted in waves or in a stream of particles—heat and light from a fire, for example, is a form of radiation.

82. **Radioactivity:** emissions—radiation, or energy waves—produced by the decay of atoms.

83. **Reductionism:** the practice of seeing something complex as merely the sum of its parts, without considering how those parts may interact with each other.

84. **ScienceDirect:** a scientific database that offers peer-reviewed journal articles and book chapters in the sphere of science and medicine.

85. **Self-regulating Earth:** the position that all organisms interact with the Earth's air, water, and rocks in order to keep the planet fit and comfortable for life in a stable fashion.

86. **Soviet Union:** also known as the Union of Soviet Socialist Republics (USSR), a

socialist state on the European continent. The USSR, in existence 1922–91, was a one-party state governed by the Communist Party. Moscow was its capital city.

87. **Space exploration:** the investigation of biophysical conditions beyond the boundary of Earth—space, stars, planets, comets, etc—using manned spacecraft, satellites or probes.

88. **Sputnik:** a Russian satellite launched into space in 1957. It was the first human-made object ever to leave the Earth's atmosphere.

89. **Superorganism:** an entity made up of many distinct organisms.

90. **Teleological:** relating to teleology—an approach that explains the reason or explanation for something in terms of its end, purpose, or goal.

91. **Theology:** the study of God and religious faiths.

92. **Theory:** a natural explanation for a group of facts or phenomena. Theories are coherent, predictive, systematic, and widely applicable. Theories generally undergo testing and improvement or modification as more information is generated so that the predictive capacity of a theory becomes greater over time.

93. **Thermostat:** an automatic apparatus for regulating temperature in an electrical device such as a kitchen oven.

94. **Ultraviolet:** part of the spectrum of light invisible to the human eye.

95. **United Nations:** an international institution founded in 1945 with the aim of promoting international cooperation, peace, and security.

96. **Viking Mission to Mars:** NASA's mission to Mars; the probes launched in 1975 were the first to land safely on the surface and return images.

97. **Wollaston medal:** the highest award in the field of geoscience given by the Geological Society of London. Geologists who have contributed significantly through excellent research in fundamental and/or applied aspects of geoscience may be honored with this award.

98. **Working class:** a socioeconomic term used to describe people in a social class marked by jobs that provide relatively low pay, require limited skill sets, and low educational requirements.

PEOPLE MENTIONED IN THE TEXT

1. **Meinrat Andreae (b. 1949)** is a German mineralogist, currently serving as the director of the biogeochemistry department at the Max Planck Institute for Chemistry in Germany.

2. **Penelope J. Boston** is a professor in the department of Earth and environmental science at New Mexico Tech University. Boston is well known for proposing that small jumping robots be sent to Mars to facilitate exploration.

3. **Robert Charlson** is a professor emeritus of atmospheric sciences and chemistry at the University ofWashington. Charlson collaborated with James Lovelock and two other scientists, Meinrat Andreae and Stephen G. Warren, on the CLAW hypothesis.

4. **Eileen Crist** is a sociologist, and an associate professor of science and technology in society at Virginia Polytechnic Institute and State University (Virginia Tech).

5. **Charles Darwin (1809–82)** was an English naturalist and geologist. Darwin is best known for his theory of evolution by means of natural selection.

6. **Richard Dawkins (b. 1941)** is a British evolutionary biologist, a popular writer, and an outspoken atheist. Dawkins served as a professor at the University of Oxford and is now an emeritus fellow of New College, Oxford.

7. **Ford Doolittle (b. 1941)** is a biochemist who was born in the US state of Illinois. He is a professor at Dalhousie University in Canada and has been a long-term critic of Lovelock's Gaia hypothesis.

8. **René Dubos (1901–82)** was a French-born American microbiologist and a professor emeritus at Rockefeller University. Dubos was one of the earlier supporters of the main ideas of the Gaia hypothesis.

9. **Nellie A. Elizabeth** was James Lovelock's mother who worked as a personal secretary. She used to take Lovelock to the local library to borrow science fiction books to read.

10. **Sidney Epton** worked for Shell'sThornton Research Centre in the United Kingdom and collaborated with James Lovelock in the 1970s in developing the Gaia hypothesis.

11. **William Golding (1911–93)** was an English novelist who the Nobel Prize for Literature in 1983. A resident of Lovelock's village, he suggested the name "Gaia" for Lovelock's hypothesis.
12. **Stephen J. Gould (1941–2002)** was an American paleontologist and evolutionary biologist who spent most of his career teaching at Harvard University. Gould was very critical of Lovelock's Gaia hypothesis.
13. **Václav Havel (1936–2011)** was a Czech playwright, president of Czechoslovakia between 1989 and 1992 and of the Czech Republic between 1993 and 2003.
14. **Dian Hitchcock** is a philosopher who worked with James Lovelock at NASA, where they examined the atmospheric data from Mars and concluded that Mars was lifeless. A decade later, it was confirmed by missions to Mars that their conclusions were correct.
15. **Heinrich Holland (1927–2012)** was born in Germany, but settled in the United States. Holland was an emeritus professor at Harvard University and he made major contributions to the understanding of the Earth's geochemistry.

16. **George Hutchinson (1903–91)** was an American zoologist, best known for his ecological studies of freshwater lakes. Hutchinson was born in England and educated at Cambridge University. In 1928, he joined the faculty of Yale University to teach zoology and he spent most of his professional life there.
17. **James Hutton (1726–97)** was a highly influential geologist from Scotland whose ideas anticipated those of Lovelock.
18. **Thomas H. Huxley (1825–95)** was an English biologist, best known for his work on Charles Darwin's theory of evolution by natural selection. Huxley did more than anyone else to get Darwin's theory accepted by scientists and the general public.
19. **John F. Kennedy (1917–63)** also known as JFK, was the 35th president of the United States (1961–63) and the youngest man elected to the office. On November 22, 1963, he was assassinated in Dallas, Texas, becoming also the youngest president to die.
20. **James W. Kirchner** is currently a professor of Earth and planetary science at

the University of California, Berkeley. Kirchner has been very interested in, if critical of, the Gaia hypothesis.

21. **Yevgraf M. Korolenko** was a nineteenth century Ukrainian philosopher and scientist. Korolenko was a learned man, although self-educated; he was familiar with the works of the great natural scientists of his time.

22. **Timothy Lenton** is professor of climate change and Earth system science at the University of Exeter. Throughout his career Lenton has been very interested in James Lovelock's controversial Gaia hypothesis and is considered a possible successor to Lovelock.

23. **Thomas A. Lovelock** was James Lovelock's father. In professional life, Tom Lovelock was an art dealer and had very strong feelings about nature.

24. **Lynn Margulis (1938–2011)** was a distinguished professor of geosciences at the University of Massachusetts, Amherst. Margulis collaborated with James Lovelock and together they developed the controversial Gaia hypothesis.

25. **Mary Midgley (b. 1919)** is an English moral philosopher whose special interest covers science, ethics, human nature, and animal rights. Midgley wrote in favor of a moral interpretation of Lovelock's Gaia hypothesis.

26. **Vance Oyama (1922–98)** was a biochemist who worked for NASA on the search for life on Mars. Oyama will be remembered for his pioneering life detection experiments on Apollo lunar samples.

27. **H. Bruce Rinker** is an ecologist and executive director of the Valley Conservation Council in Virginia.

28. **Peter Saunders** is emeritus professor of mathematics at King's College, London.

29. **Stephen Schneider (1945–2010)** was an American professor of environmental biology and global change at Stanford University. He was internationally recognized for his research, policy analysis, and outreach in the subject of global climate change.

30. **Eduard Suess (1831–1914)** was an Austrian geologist who contributed greatly to the knowledge of his field. Suess is credited with generating many of the

concepts that led to the theory of plate tectonics (the movement of the Earth's crust) and paleogeography (the study of ancient land masses).

31. **Toby Tyrrell** is a professor of Earth system science at the University of Southampton. Tyrrell is well known for his critical review of the Gaia hypothesis.

32. **Vladimir I. Vernadsky (1863–1945)** was a renowned Russian mineralogist and geochemist. Vernadsky is considered the founder and father of modern geochemistry.

33. **Jules Verne (1828–1905)** was a nineteenth-century French novelist and poet. He is the author of *Around the World in Eighty Days* and *20,000 Leagues Under the Sea*.

34. **Peter Ward (b. 1949)** is an American paleontologist and currently a professor at the University of Adelaide in Australia. He is known for his anti-Gaian hypothesis, *The Medea Hypothesis*, published in 2009.

35. **Andrew Watson (b. 1952)** is a British marine and atmospheric scientist, currently a professor at the University of Exeter. Watson was a PhD student of James Lovelock; together, they developed the computer model "Daisyworld" to provide proof that organisms can regulate their environment.

36. **Alfred Wegener (1880–1930)** was a polar researcher, geophysicist, and meteorologist born in Berlin, Germany. Wegener became famous for his groundbreaking theory of continental drift that, for the first time, suggested that all continents were slowly moving around the Earth.

37. **H. G. Wells (1866–1946)** was a British novelist. The author of *The Time Machine* and *The War of the Worlds*, Wells is best known for his science fiction novels.

WORKS CITED

1. Ayers, Greg and Jill Cainey. "The CLAW Hypothesis: A Review of the Major Developments." *Environmental Chemistry* 4 (2007): 366–74.
2. Bauman, Brent F. "The Feasibility of a Testable Gaia Hypothesis." BSc Thesis, James Madison University, 1998.
3. Carmo, Ricardo S. do, Nei Freitas Nunes-Neto, and Charbel Nino El-Hani. "Gaia Theory in Brazilian High School Biology Textbooks." *Science & Education* 18 (2009): 469–501.
4. Crist, Eileen, and H. Bruce Rinker. "One Grand Organic Whole." In *Gaia in Turmoil: Climate Change, Biodepletion, and Earth Ethics in an Age of Crisis*, edited by Eileen Crist and H. Bruce Rinker, 3–20. Cambridge: The MIT Press, 2010.
5. Darwin, Charles. *On the Origin of Species by Means of Natural Selection, or the Preservation of Favoured Races in the Struggle for Life*. London: John Murray, 1859.
6. Davis, David. "A Few Thoughts on the Apocalypse." *The Spectator*, February 25, 2009. Accessed December 21, 2013. http://www.spectator.co.uk/features/3387731/a-few-thoughts-on-the-apocalypse/.
7. Dawkins, Richard. *The Extended Phenotype: The Gene as the Unit of Selection*. Oxford: Oxford University Press, 1982.
8. Doolittle, W. Ford. "Is Nature Really Motherly?" *CoEvolution Quarterly* 29 (1981): 58–65.
9. Dubos, René. "Gaia and Creative Evolution." *Nature* 282 (1979): 154–55.
10. Environment website. "Gaia Hypothesis." Accessed December 23, 2013. http://www.environment.gen.tr/gaia/70-gaia-hypothesis.html.
11. Flint, James. "Earth—The Final Conflict." The *Daily Telegraph*, February 6, 2006. Accessed December 23, 2013. http://www.telegraph.co.uk/culture/books/3649909/Earth-the-final-conflict.html.
12. Forbes, Peter. "Jim'll Fix It." The *Guardian*, February 21, 2009. Accessed December 23, 2013. http://www.theguardian.com/culture/2009/feb/21/james-lovelock-gaia-book-review.
13. Gaia Theory Conference. "Gaia Theory Conference at George Mason University." Arlington County. Accessed December 27, 2013. http://www.gaiatheory. org/2006-conference/.

14. Goodell, Jeff. "James Lovelock, the Prophet." *Rolling Stone*, November 1, 2007. Accessed December 23, 2013. http://www.rollingstone.com/politics/news/ james-lovelock-the-prophet-20071101.

15. Gould, Stephen J. "Kropotkin Was No Crackpot." *Natural History* 106 (1997): 12-21.

16. Gray, John. "James Lovelock: A Man for All Seasons." *New Statesman*, March 27, 2013. Accessed December 21, 2013. http://www.newstatesman.com/ culture/ culture/2013/03/james-lovelock-man-all-seasons.

17. "The Revenge of Gaia, by James Lovelock." The *Independent*, January 27, 2006. Accessed December 24, 2013. http://www.independent.co.uk/arts-entertainment/ books/reviews/the-revenge-of-gaia-by-james-lovelock-6110631. htmlhttp://www. independent.co.uk/arts-entertainment/books/reviews/the-revenge-of-gaia-by-james-lovelock-524635.html.

18. Hauk, Marna, Judith Landsman, Jeanine M. Canty, and Noël Cox Caniglia. "Gaian Methodologies: An Emergent Confluence of Sustainability Research Innovation." Paper presented at the Association for the Advancement of Sustainability in Higher Education Conference, Denver, October 10–12, 2010.

19. Irvine, Ian. "James Lovelock: The Green Man." The *Independent*, December 3, 2005. Accessed October 10, 2013. http://www.independent.co.uk/news/people/ profiles/james-lovelock-the-green-man-517953.html.

20. James Lovelock's official website. "Curriculum Vitae." Accessed December 29, 2013. http://www.jameslovelock.org/page2.html.

21. Kasting, James F. "The Gaia Hypothesis Is Still Giving Us Feedback." *Nautilus* 12 (2014).

22. Kauffman, Eric G. "The Gaia Controversy: AGU's Chapman Conference." *Eos, Transactions of the American Geophysical Union* 69, no. 31 (1988): 763–64.

23. Kennedy, John F., Presidential Library and Museum. "Space Program." Accessed January 8, 2016. http://www.jfklibrary.org/JFK/JFK-in-History/Space-Program. aspx.

24. Kirchner, James W. "Gaia Hypothesis: Can it Be Tested?" *Reviews of Geophysics* 27, no. 2 (1989): 223–35.

25. "The Gaia Hypothesis: Conjectures and Refutations." *Climatic Change* 58 (2003): 21–45.

26. "The Gaia Hypothesis: Fact, Theory, and Wishful Thinking." *Climatic* Change 52 (2002): 391–408.

27. Lenton, Timothy M. "Gaia and Natural Selection." *Nature* 394 (1998): 439–47.

28. Lenton, Timothy M., and Andrew Watson. *Revolutions that Made the Earth.* Oxford: Oxford University Press, 2013.

29. Lovelock, James E. *The Ages of Gaia: A Biography of our Living Earth*, Rev. ed. Oxford: Oxford University Press, 1995.

30. *Gaia: A New Look at Life on Earth*. Rev. ed. Oxford: Oxford University Press, 2000.

31. *Gaia: Medicine for an Ailing Planet*. London: Gaia Books, 2005.

32. "Geophysiology, the Science of Gaia." *Reviews of Geophysics* 27 (1989): 215–22.

33. *Healing Gaia: The Practical Science of Planetary Medicine*. Oxford: Oxford University Press, 2001.

34. *Homage to Gaia: The Life of an Independent Scientist*, Rev. ed. London: Souvenir Press Ltd., 2014.

35. "Gaia: The Living Earth." *Nature* 426 (2003): 769–70.

36. *The Revenge of Gaia: Why the Earth is Fighting Back—and How we Can Still Save Humanity*. London: Penguin Books, 2007.

37. *A Rough Ride to the Future*. London: Penguin Books, 2015.

38. *The Vanishing Face of Gaia: A Final Warning.* London: Penguin Books, 2010.

39. "Wollaston Medal Citation." Accessed March 4, 2016. http://www. jameslovelock.org/page7.html.

40. Lovelock, James E., and Sidney Epton. "The Quest for Gaia." *New Scientist* 65, no. 935 (1975): 304–09.

41. Lovelock, James E., and C. E. Giffin. "Planetary Atmospheres: Compositional and Other Changes Associated with the Presence of Life." *Advances in the Astronautical Sciences* 25 (1969): 179–93.

42. Lovelock, James E., and Lynn Margulis. "Atmospheric Homeostasis by and for the Biosphere: The Gaia Hypothesis." *Tellus* 26, nos. 1–2 (1974): 2–10.

43. McKie, Robin. "Gaia's Warrior." *Green Lifestyle Magazine*, July/August 2007.

44. Mellanby, Kenneth. "Living with the Earth Mother." *New Scientist* 84 (1979): 41.

45. Midgley, Mary. "Great Thinkers—James Lovelock." *New Statesman*, 14 July 2003.

46. Ogle, Martin. "The Gaia Theory: Scientific Model and Metaphor for the 21st Century." *Revista Umbral* (*Threshold Magazine*) 1 (2009): 99–106.

47. Ravilious, Kate. "Perfect Harmony." The *Guardian*, April 28, 2008. Accessed December 30, 2013. www.theguardian.com/science/2008/apr/28/scienceofclimatechange.biodiversity.

48. Schneider, Stephen H., and Penelope J. Boston, eds. *Scientists on Gaia*. Cambridge: The MIT Press, 1993.

49. Schneider, Stephen H., James R. Miller, Eileen Crist, and Penelope J. Boston, eds. *Scientists Debate Gaia: The Next Century*. Cambridge: The MIT Press, 2004.

50. Tickell, Crispin. "Scientists on Gaia." The *Financial Times* 2002. Accessed December 23, 2013. http://www.crispintickell.com/page19.html.

51. Turney, Jon. *Lovelock and Gaia: Signs of Life*. New York: Columbia University Press, 2003.

52. Tyrrell, Toby. *On Gaia: A Critical Investigation of the Relationship Between Life and Earth*. Princeton: Princeton University Press, 2013.

53. Velasquez-Manoff, Moises. "The Medea Hypothesis: A Response to the Gaia Hypothesis." *The Christian Science Monitor*, February 12, 2010. Accessed December 28, 2013. http://www.csmonitor.com/Environment/Bright-Green/2010/0212/The-Medea-Hypothesis-A-response-to-the-Gaia-hypothesis.

54. Wallace, Richard R., and Bryan G. Norton. "Policy Implications of Gaian Theory." *Ecological Economics* 6 (1992): 103–18.

55. Ward, Peter. *The Medea Hypothesis: Is Life on Earth Ultimately Self-Destructive?* Princeton: Princeton University Press, 2009.

56. Watson, Andrew J., and James E. Lovelock. "Biological Homeostasis of the Global Environment: The Parable of Daisyworld." *Tellus* 35B (1983): 286–89.

原书作者简介

詹姆斯·E. 拉伍洛克，1919 年出生于英格兰一个工人阶级家庭，父母几乎没有受过教育。尽管早年生活困顿，但他获得了多个学位。之后，拉伍洛克发明了若干科学仪器。这从经济上帮助他在大学系统之外从事研究。在美国国家航空航天局工作期间，他对生命和环境的相互作用感到着迷。1979 年，他的第一部作品《盖娅：地球生命的新视野》提出了具有争议的论点，即我们的星球进行着自我管理。有关拉伍洛克"盖娅假说"的辩论持续了数十年。他一直坚持扩充并为其作品辩护，还于 2015 年以 95 岁高龄发表了第十部盖娅的相关作品。

本书作者简介

穆罕默德·沙姆斯乌杜哈博士，伦敦大学学院水文地质学博士。现为伦敦大学学院风险与减灾研究院研究员。

世界名著中的批判性思维

《世界思想宝库钥匙丛书》致力于深入浅出地阐释全世界著名思想家的观点，不论是谁、在何处都能了解到，从而推进批判性思维发展。

《世界思想宝库钥匙丛书》与世界顶尖大学的一流学者合作，为一系列学科中最有影响的著作推出新的分析文本，介绍其观点和影响。在这一不断扩展的系列中，每种选入的著作都代表了历经时间考验的思想典范。通过为这些著作提供必要背景、揭示原作者的学术渊源以及说明这些著作所产生的影响，本系列图书希望让读者以新视角看待这些划时代的经典之作。读者应学会思考、运用并挑战这些著作中的观点，而不是简单接受它们。

ABOUT THE AUTHOR OF THE ORIGINAL WORK

James Lovelock was born in England in 1919, to working-class parents with little education. Despite financial struggles early in life, |he earned several degrees. Lovelock then invented a number of scientific instruments, which gave him the financial means to do research outside the university system.While working with NASA (the National Aeronautics and Space Administration) he grew fascinated by how life and the environment interact. In 1979, his first book Gaia: A New Look at Life on Earth put forward the controversial idea that our planet is self-regulating. The debate about Lovelock's 'Gaia hypothesis' has simmered for decades. He continues to expand and defend his work, and in 2015 published his tenth book on Gaia, at the age of 95.

ABOUT THE AUTHORS OF THE ANALYSIS

Dr Mohammad Shamsudduha holds a PhD in hydrogeology from University College, London. He is currently a researcher at the University College, London, Institute for Risk and Disaster Reduction.

ABOUT MACAT
GREAT WORKS FOR CRITICAL THINKING

Macat is focused on making the ideas of the world's great thinkers accessible and comprehensible to everybody, everywhere, in ways that promote the development of enhanced critical thinking skills.

It works with leading academics from the world's top universities to produce new analyses that focus on the ideas and the impact of the most influential works ever written across a wide variety of academic disciplines. Each of the works that sit at the heart of its growing library is an enduring example of great thinking. But by setting them in context — and looking at the influences that shaped their authors, as well as the responses they provoked — Macat encourages readers to look at these classics and game-changers with fresh eyes. Readers learn to think, engage and challenge their ideas, rather than simply accepting them.

批判性思维与《盖娅：地球生命的新视野》

首要批判性思维技巧：阐释

次要批判性思维技巧：理性化思维

《盖娅：地球生命的新视野》一书可能会继续导致意见分歧，但没有人可以否认这部作品有助于深入理解其作者詹姆斯·E.拉伍洛克的创新思维。

《盖娅》于1979年出版，书中提出了一种全新的假设：拉伍洛克认为地球是一个有生命的实体。这座行星和它全部的独立生物体组成了一个可自我调节的整体，维持生命并跟随时间不断进化。拉伍洛克眼中的人类，与地球上的其他组成部分并无不同，他反对人类的利益高于一切这一曾经被广泛认可的观点。

尽管发表之初人们认为拉伍洛克的观点激进甚至愚蠢，但对其视角的某一种解读却在当代有关环境与气候的辩论中引发共鸣，并且如今得到现代思想家们广泛接受。随着人类对气候的影响愈发极端，地球越来越多的自我调节特性被揭示出来。于是，科学家们不得不好奇这颗行星为了有效自我调节究竟可以做到什么程度。诚然，虽然拉伍洛克的盖娅理论存在牵强附会的地方，但今天读来比过去以往都更令人信服。通过批判性思维技巧而得出的结论，让拉伍洛克能够为现有证据创造出全新的解释，尤其是运用新的形式把碎片状的现有证据拼接为一个整体。

CRITICAL THINKING AND *GAIA: A NEW LOOK AT LIFE ON EARTH*

- Primary critical thinking skill: INTERPRETATION
- Secondary critical thinking skill: REASONING

Gaia: A New Look At Life on Earth may continue to divide opinion, but nobody can deny that the book offers a powerful insight into the creative thinking of its author, James E. Lovelock.

Published in 1979, *Gaia* offered a radically new hypothesis: the Earth, Lovelock argued, is a living entity. Together, the planet and all its separate living organisms form a single self-regulating body, sustaining life and helping it evolve through time. Lovelock sees humans as no more special than other elements of the planet, railing against the once widely-held belief that the good of mankind is the only thing that matters.

Despite being seen as radical, and even idiotic on its publication, a version of Lovelock's viewpoint has found resonance in contemporary debates about the environment and climate, and has now broadly come to be accepted by modern thinkers. As man's effects on the climate become increasingly extreme, more and more elements of the Earth's self-regulation seem to be unveiled—forcing scientists to ask how far the planet might be able to go in order self-regulate effectively. Indeed, despite its far-fetched elements, Lovelock's Gaia thesis seems to ring more convincingly today than ever before; that it does is largely a result of the critical thinking skills that allowed Lovelock to produce novel explanations for existing evidence and, above all, to connect existing fragments of evidence together in new ways.

《世界思想宝库钥匙丛书》简介

《世界思想宝库钥匙丛书》致力于为一系列在各领域产生重大影响的人文社科类经典著作提供独特的学术探讨。每一本读物都不仅仅是原经典著作的内容摘要，而是介绍并深入研究原经典著作的学术渊源、主要观点和历史影响。这一丛书的目的是提供一套学习资料，以促进读者掌握批判性思维，从而更全面、深刻地去理解重要思想。

每一本读物分为 3 个部分：学术渊源、学术思想和学术影响，每个部分下有 4 个小节。这些章节旨在从各个方面研究原经典著作及其反响。

由于独特的体例，每一本读物不但易于阅读，而且另有一项优点：所有读物的编排体例相同，读者在进行某个知识层面的调查或研究时可交叉参阅多本该丛书中的相关读物，从而开启跨领域研究的路径。

为了方便阅读，每本读物最后还列出了术语表和人名表（在书中则以星号 * 标记），此外还有参考文献。

《世界思想宝库钥匙丛书》与剑桥大学合作，理清了批判性思维的要点，即如何通过 6 种技能来进行有效思考。其中 3 种技能让我们能够理解问题，另 3 种技能让我们有能力解决问题。这 6 种技能合称为“批判性思维 PACIER 模式”，它们是：

分析：了解如何建立一个观点；
评估：研究一个观点的优点和缺点；
阐释：对意义所产生的问题加以理解；
创造性思维：提出新的见解，发现新的联系；
解决问题：提出切实有效的解决办法；
理性化思维：创建有说服力的观点。

THE MACAT LIBRARY

The Macat Library is a series of unique academic explorations of seminal works in the humanities and social sciences — books and papers that have had a significant and widely recognised impact on their disciplines. It has been created to serve as much more than just a summary of what lies between the covers of a great book. It illuminates and explores the influences on, ideas of, and impact of that book. Our goal is to offer a learning resource that encourages critical thinking and fosters a better, deeper understanding of important ideas.

Each publication is divided into three Sections: Influences, Ideas, and Impact. Each Section has four Modules. These explore every important facet of the work, and the responses to it.

This Section-Module structure makes a Macat Library book easy to use, but it has another important feature. Because each Macat book is written to the same format, it is possible (and encouraged!) to cross-reference multiple Macat books along the same lines of inquiry or research. This allows the reader to open up interesting interdisciplinary pathways.

To further aid your reading, lists of glossary terms and people mentioned are included at the end of this book (these are indicated by an asterisk [*] throughout) — as well as a list of works cited.

Macat has worked with the University of Cambridge to identify the elements of critical thinking and understand the ways in which six different skills combine to enable effective thinking.

Three allow us to fully understand a problem; three more give us the tools to solve it. Together, these six skills make up the PACIER model of critical thinking. They are:

ANALYSIS — understanding how an argument is built
EVALUATION — exploring the strengths and weaknesses of an argument
INTERPRETATION — understanding issues of meaning
CREATIVE THINKING — coming up with new ideas and fresh connections
PROBLEM-SOLVING — producing strong solutions
REASONING — creating strong arguments

“《世界思想宝库钥匙丛书》提供了独一无二的跨学科学习和研究工具。它介绍那些革新了各自学科研究的经典著作，还邀请全世界一流专家和教育机构进行严谨的分析，为每位读者打开世界顶级教育的大门。”

—— 安德烈亚斯·施莱歇尔，
经济合作与发展组织教育与技能司司长

“《世界思想宝库钥匙丛书》直面大学教育的巨大挑战……他们组建了一支精干而活跃的学者队伍，来推出在研究广度上颇具新意的教学材料。”

—— 布罗尔斯教授、勋爵，剑桥大学前校长

“《世界思想宝库钥匙丛书》的愿景令人赞叹。它通过分析和阐释那些曾深刻影响人类思想以及社会、经济发展的经典文本，提供了新的学习方法。它推动批判性思维，这对于任何社会和经济体来说都是至关重要的。这就是未来的学习方法。”

—— 查尔斯·克拉克阁下，英国前教育大臣

“对于那些影响了各自领域的著作，《世界思想宝库钥匙丛书》能让人们立即了解到围绕那些著作展开的评论性言论，这让该系列图书成为在这些领域从事研究的师生们不可或缺的资源。”

—— 威廉·特朗佐教授，加利福尼亚大学圣地亚哥分校

"Macat offers an amazing first-of-its-kind tool for interdisciplinary learning and research. Its focus on works that transformed their disciplines and its rigorous approach, drawing on the world's leading experts and educational institutions, opens up a world-class education to anyone."

—— Andreas Schleicher, Director for Education and Skills, Organisation for Economic Co-operation and Development

"Macat is taking on some of the major challenges in university education... They have drawn together a strong team of active academics who are producing teaching materials that are novel in the breadth of their approach."

—— Prof Lord Broers, former Vice-Chancellor of the University of Cambridge

"The Macat vision is exceptionally exciting. It focuses upon new modes of learning which analyse and explain seminal texts which have profoundly influenced world thinking and so social and economic development. It promotes the kind of critical thinking which is essential for any society and economy. This is the learning of the future."

—— Rt Hon Charles Clarke, former UK Secretary of State for Education

"The Macat analyses provide immediate access to the critical conversation surrounding the books that have shaped their respective discipline, which will make them an invaluable resource to all of those, students and teachers, working in the field."

—— Prof William Tronzo, University of California at San Diego

The Macat Library
世界思想宝库钥匙丛书

TITLE	中文书名	类别
An Analysis of Arjun Appadurai's *Modernity at Large: Cultural Dimensions of Globalization*	解析阿尔君·阿帕杜莱《消失的现代性：全球化的文化维度》	人类学
An Analysis of Claude Lévi-Strauss's *Structural Anthropology*	解析克劳德·列维–施特劳斯《结构人类学》	人类学
An Analysis of Marcel Mauss's *The Gift*	解析马塞尔·莫斯《礼物》	人类学
An Analysis of Jared M. Diamond's *Guns, Germs, and Steel: The Fate of Human Societies*	解析贾雷德·戴蒙德《枪炮、病菌与钢铁：人类社会的命运》	人类学
An Analysis of Clifford Geertz's *The Interpretation of Cultures*	解析克利福德·格尔茨《文化的解释》	人类学
An Analysis of Philippe Ariès's *Centuries of Childhood: A Social History of Family Life*	解析菲力浦·阿利埃斯《儿童的世纪：旧制度下的儿童和家庭生活》	人类学
An Analysis of W. Chan Kim & Renée Mauborgne's *Blue Ocean Strategy*	解析金伟灿 / 勒妮·莫博涅《蓝海战略》	商业
An Analysis of John P. Kotter's *Leading Change*	解析约翰·P. 科特《领导变革》	商业
An Analysis of Michael E. Porter's *Competitive Strategy: Techniques for Analyzing Industries and Competitors*	解析迈克尔·E. 波特《竞争战略：分析产业和竞争对手的技术》	商业
An Analysis of Jean Lave & Etienne Wenger's *Situated Learning: Legitimate Peripheral Participation*	解析琼·莱夫 / 艾蒂纳·温格《情境学习：合法的边缘性参与》	商业
An Analysis of Douglas McGregor's *The Human Side of Enterprise*	解析道格拉斯·麦格雷戈《企业的人性面》	商业
An Analysis of Milton Friedman's *Capitalism and Freedom*	解析米尔顿·弗里德曼《资本主义与自由》	商业
An Analysis of Ludwig von Mises's *The Theory of Money and Credit*	解析路德维希·冯·米塞斯《货币和信用理论》	经济学
An Analysis of Adam Smith's *The Wealth of Nations*	解析亚当·斯密《国富论》	经济学
An Analysis of Thomas Piketty's *Capital in the Twenty-First Century*	解析托马斯·皮凯蒂《21 世纪资本论》	经济学
An Analysis of Nassim Nicholas Taleb's *The Black Swan: The Impact of the Highly Improbable*	解析纳西姆·尼古拉斯·塔勒布《黑天鹅：如何应对不可预知的未来》	经济学
An Analysis of Ha-Joon Chang's *Kicking Away the Ladder*	解析张夏准《富国陷阱：发达国家为何踢开梯子》	经济学
An Analysis of Thomas Robert Malthus's *An Essay on the Principle of Population*	解析托马斯·罗伯特·马尔萨斯《人口论》	经济学

An Analysis of John Maynard Keynes's *The General Theory of Employment, Interest and Money*	解析约翰·梅纳德·凯恩斯《就业、利息和货币通论》	经济学
An Analysis of Milton Friedman's *The Role of Monetary Policy*	解析米尔顿·弗里德曼《货币政策的作用》	经济学
An Analysis of Burton G. Malkiel's *A Random Walk Down Wall Street*	解析伯顿·G. 马尔基尔《漫步华尔街》	经济学
An Analysis of Friedrich A. Hayek's *The Road to Serfdom*	解析弗里德里希·A. 哈耶克《通往奴役之路》	经济学
An Analysis of Charles P. Kindleberger's *Manias, Panics, and Crashes: A History of Financial Crises*	解析查尔斯·P. 金德尔伯格《疯狂、惊恐和崩溃：金融危机史》	经济学
An Analysis of Amartya Sen's *Development as Freedom*	解析阿马蒂亚·森《以自由看待发展》	经济学
An Analysis of Rachel Carson's *Silent Spring*	解析蕾切尔·卡森《寂静的春天》	地理学
An Analysis of Charles Darwin's *On the Origin of Species: by Means of Natural Selection, or The Preservation of Favoured Races in the Struggle for Life*	解析查尔斯·达尔文《物种起源》	地理学
An Analysis of World Commission on Environment and Development's *The Brundtland Report, Our Common Future*	解析世界环境与发展委员会《布伦特兰报告：我们共同的未来》	地理学
An Analysis of James E. Lovelock's *Gaia: A New Look at Life on Earth*	解析詹姆斯·E. 拉伍洛克《盖娅：地球生命的新视野》	地理学
An Analysis of Paul Kennedy's *The Rise and Fall of the Great Powers: Economic Change and Military Conflict from 1500—2000*	解析保罗·肯尼迪《大国的兴衰：1500—2000 年的经济变革与军事冲突》	历史
An Analysis of Janet L. Abu-Lughod's *Before European Hegemony: The World System A. D. 1250—1350*	解析珍妮特·L. 阿布–卢格霍德《欧洲霸权之前：1250—1350 年的世界体系》	历史
An Analysis of Alfred W. Crosby's *The Columbian Exchange: Biological and Cultural Consequences of 1492*	解析艾尔弗雷德·W. 克罗斯比《哥伦布大交换：1492 年以后的生物影响和文化冲击》	历史
An Analysis of Tony Judt's *Postwar: A History of Europe since 1945*	解析托尼·朱特《战后欧洲史》	历史
An Analysis of Richard J. Evans's *In Defence of History*	解析理查德·J. 艾文斯《捍卫历史》	历史
An Analysis of Eric Hobsbawm's *The Age of Revolution: Europe 1789–1848*	解析艾瑞克·霍布斯鲍姆《革命的年代：欧洲 1789—1848 年》	历史

An Analysis of Roland Barthes's *Mythologies*	解析罗兰·巴特《神话学》	文学与批判理论
An Analysis of Simone de Beauvoir's *The Second Sex*	解析西蒙娜·德·波伏娃《第二性》	文学与批判理论
An Analysis of Edward W. Said's *Orientalism*	解析爱德华·W. 萨义德《东方主义》	文学与批判理论
An Analysis of Virginia Woolf's *A Room of One's Own*	解析弗吉尼亚·伍尔芙《一间自己的房间》	文学与批判理论
An Analysis of Judith Butler's *Gender Trouble*	解析朱迪斯·巴特勒《性别麻烦》	文学与批判理论
An Analysis of Ferdinand de Saussure's *Course in General Linguistics*	解析费尔迪南·德·索绪尔《普通语言学教程》	文学与批判理论
An Analysis of Susan Sontag's *On Photography*	解析苏珊·桑塔格《论摄影》	文学与批判理论
An Analysis of Walter Benjamin's *The Work of Art in the Age of Mechanical Reproduction*	解析瓦尔特·本雅明《机械复制时代的艺术作品》	文学与批判理论
An Analysis of W.E.B. Du Bois's *The Souls of Black Folk*	解析 W.E.B. 杜波依斯《黑人的灵魂》	文学与批判理论
An Analysis of Plato's *The Republic*	解析柏拉图《理想国》	哲学
An Analysis of Plato's *Symposium*	解析柏拉图《会饮篇》	哲学
An Analysis of Aristotle's *Metaphysics*	解析亚里士多德《形而上学》	哲学
An Analysis of Aristotle's *Nicomachean Ethics*	解析亚里士多德《尼各马可伦理学》	哲学
An Analysis of Immanuel Kant's *Critique of Pure Reason*	解析伊曼努尔·康德《纯粹理性批判》	哲学
An Analysis of Ludwig Wittgenstein's *Philosophical Investigations*	解析路德维希·维特根斯坦《哲学研究》	哲学
An Analysis of G.W.F. Hegel's *Phenomenology of Spirit*	解析 G.W.F. 黑格尔《精神现象学》	哲学
An Analysis of Baruch Spinoza's *Ethics*	解析巴鲁赫·斯宾诺莎《伦理学》	哲学
An Analysis of Hannah Arendt's *The Human Condition*	解析汉娜·阿伦特《人的境况》	哲学
An Analysis of G.E.M. Anscombe's *Modern Moral Philosophy*	解析 G.E.M. 安斯康姆《现代道德哲学》	哲学
An Analysis of David Hume's *An Enquiry Concerning Human Understanding*	解析大卫·休谟《人类理解研究》	哲学

An Analysis of Søren Kierkegaard's *Fear and Trembling*	解析索伦·克尔凯郭尔《恐惧与战栗》	哲学
An Analysis of René Descartes's *Meditations on First Philosophy*	解析勒内·笛卡尔《第一哲学沉思录》	哲学
An Analysis of Friedrich Nietzsche's *On the Genealogy of Morality*	解析弗里德里希·尼采《论道德的谱系》	哲学
An Analysis of Gilbert Ryle's *The Concept of Mind*	解析吉尔伯特·赖尔《心的概念》	哲学
An Analysis of Thomas Kuhn's *The Structure of Scientific Revolutions*	解析托马斯·库恩《科学革命的结构》	哲学
An Analysis of John Stuart Mill's *Utilitarianism*	解析约翰·斯图亚特·穆勒《功利主义》	哲学
An Analysis of Aristotle's *Politics*	解析亚里士多德《政治学》	政治学
An Analysis of Niccolò Machiavelli's *The Prince*	解析尼科洛·马基雅维利《君主论》	政治学
An Analysis of Karl Marx's *Capital*	解析卡尔·马克思《资本论》	政治学
An Analysis of Benedict Anderson's *Imagined Communities*	解析本尼迪克特·安德森《想象的共同体》	政治学
An Analysis of Samuel P. Huntington's *The Clash of Civilizations and the Remaking of World Order*	解析塞缪尔·P. 亨廷顿《文明的冲突与世界秩序的重建》	政治学
An Analysis of Alexis de Tocqueville's *Democracy in America*	解析阿列克西·德·托克维尔《论美国的民主》	政治学
An Analysis of John A. Hobson's *Imperialism: A Study*	解析约翰·A. 霍布森《帝国主义》	政治学
An Analysis of Thomas Paine's *Common Sense*	解析托马斯·潘恩《常识》	政治学
An Analysis of John Rawls's *A Theory of Justice*	解析约翰·罗尔斯《正义论》	政治学
An Analysis of Francis Fukuyama's *The End of History and the Last Man*	解析弗朗西斯·福山《历史的终结与最后的人》	政治学
An Analysis of John Locke's *Two Treatises of Government*	解析约翰·洛克《政府论》	政治学
An Analysis of Sun Tzu's *The Art of War*	解析孙武《孙子兵法》	政治学
An Analysis of Henry Kissinger's *World Order: Reflections on the Character of Nations and the Course of History*	解析亨利·基辛格《世界秩序》	政治学
An Analysis of Jean-Jacques Rousseau's *The Social Contract*	解析让-雅克·卢梭《社会契约论》	政治学

An Analysis of Odd Arne Westad's *The Global Cold War: Third World Interventions and the Making of Our Times*	解析文安立《全球冷战：美苏对第三世界的干涉与当代世界的形成》	政治学
An Analysis of Sigmund Freud's *The Interpretation of Dreams*	解析西格蒙德·弗洛伊德《梦的解析》	心理学
An Analysis of William James' *The Principles of Psychology*	解析威廉·詹姆斯《心理学原理》	心理学
An Analysis of Philip Zimbardo's *The Lucifer Effect*	解析菲利普·津巴多《路西法效应》	心理学
An Analysis of Leon Festinger's *A Theory of Cognitive Dissonance*	解析利昂·费斯汀格《认知失调论》	心理学
An Analysis of Richard H. Thaler & Cass R. Sunstein's *Nudge: Improving Decisions about Health, Wealth, and Happiness*	解析理查德·H. 泰勒 / 卡斯·R. 桑斯坦《助推：如何做出有关健康、财富和幸福的更优决策》	心理学
An Analysis of Gordon Allport's *The Nature of Prejudice*	解析高尔登·奥尔波特《偏见的本质》	心理学
An Analysis of Steven Pinker's *The Better Angels of Our Nature: Why Violence Has Declined*	解析斯蒂芬·平克《人性中的善良天使：暴力为什么会减少》	心理学
An Analysis of Stanley Milgram's *Obedience to Authority*	解析斯坦利·米尔格拉姆《对权威的服从》	心理学
An Analysis of Betty Friedan's *The Feminine Mystique*	解析贝蒂·弗里丹《女性的奥秘》	心理学
An Analysis of David Riesman's *The Lonely Crowd: A Study of the Changing American Character*	解析大卫·理斯曼《孤独的人群：美国人社会性格演变之研究》	社会学
An Analysis of Franz Boas's *Race, Language and Culture*	解析弗朗兹·博厄斯《种族、语言与文化》	社会学
An Analysis of Pierre Bourdieu's *Outline of a Theory of Practice*	解析皮埃尔·布尔迪厄《实践理论大纲》	社会学
An Analysis of Max Weber's *The Protestant Ethic and the Spirit of Capitalism*	解析马克斯·韦伯《新教伦理与资本主义精神》	社会学
An Analysis of Jane Jacobs's *The Death and Life of Great American Cities*	解析简·雅各布斯《美国大城市的死与生》	社会学
An Analysis of C. Wright Mills's *The Sociological Imagination*	解析 C. 赖特·米尔斯《社会学的想象力》	社会学
An Analysis of Robert E. Lucas Jr.'s *Why Doesn't Capital Flow from Rich to Poor Countries?*	解析小罗伯特·E. 卢卡斯《为何资本不从富国流向穷国？》	社会学

An Analysis of Émile Durkheim's *On Suicide*	解析埃米尔·迪尔凯姆《自杀论》	社会学
An Analysis of Eric Hoffer's *The True Believer: Thoughts on the Nature of Mass Movements*	解析埃里克·霍弗《狂热分子：群众运动圣经》	社会学
An Analysis of Jared M. Diamond's *Collapse: How Societies Choose to Fail or Survive*	解析贾雷德·M. 戴蒙德《大崩溃：社会如何选择兴亡》	社会学
An Analysis of Michel Foucault's *The History of Sexuality Vol. 1: The Will to Knowledge*	解析米歇尔·福柯《性史（第一卷）：求知意志》	社会学
An Analysis of Michel Foucault's *Discipline and Punish*	解析米歇尔·福柯《规训与惩罚》	社会学
An Analysis of Richard Dawkins's *The Selfish Gene*	解析理查德·道金斯《自私的基因》	社会学
An Analysis of Antonio Gramsci's *Prison Notebooks*	解析安东尼奥·葛兰西《狱中札记》	社会学
An Analysis of Augustine's *Confessions*	解析奥古斯丁《忏悔录》	神学
An Analysis of C. S. Lewis's *The Abolition of Man*	解析 C. S. 路易斯《人之废》	神学

图书在版编目（CIP）数据

解析詹姆斯·E. 拉伍洛克《盖娅：地球生命的新视野》：汉、英/穆罕默德·沙姆斯乌杜哈（Mohammad Shamsudduha）著；庄稼译. —上海：上海外语教育出版社，2020
（世界思想宝库钥匙丛书）
ISBN 978-7-5446-6119-5

Ⅰ.①解… Ⅱ.①穆… ②庄… Ⅲ.①环境科学－哲学－研究－汉、英 Ⅳ.①X-02

中国版本图书馆CIP数据核字（2020）第011620号

This Chinese-English bilingual edition of *An Analysis of James E. Lovelock's* Gaia: A New Look at Life on Earth is published by arrangement with Macat International Limited.
Licensed for sale throughout the world.

本书汉英双语版由Macat国际有限公司授权上海外语教育出版社有限公司出版。
供在全世界范围内发行、销售。

图字：09－2018－549

出版发行：上海外语教育出版社
（上海外国语大学内） 邮编：200083
电　　话：021-65425300（总机）
电子邮箱：bookinfo@sflep.com.cn
网　　址：http://www.sflep.com
责任编辑：董　新

印　　刷：上海信老印刷厂
开　　本：890×1240　1/32　印张 6.25　字数 128千字
版　　次：2020 年 5月第 1版　2020 年 5月第 1次印刷
印　　数：2 100 册

书　　号：ISBN 978-7-5446-6119-5
定　　价：30.00 元
本版图书如有印装质量问题，可向本社调换
质量服务热线：4008-213-263　电子邮箱：editorial@sflep.com